SEX BY NUMBERS

DAVID SPIEGELHALTER is Winton Professor for the Public Understanding of Risk in the Statistical Laboratory, University of Cambridge. He is a Fellow of the Royal Society, Fellow of Churchill College Cambridge, has an OBE and knighthood for services to statistics, and in 2011 came seventh in an episode of *Winter Wipeout*. He is co-author of *The Norm Chronicles* (9781846686214), also published by Profile Books.

WELLCOME COLLECTION is the free visitor destination for the incurably curious. It explores the connections between medicine, life and art in the past, present and future. Wellcome Collection is part of the Wellcome Trust, a global charitable foundation dedicated to improving health by supporting bright minds in science, the humanities and social sciences, and public engagement.

ALSO BY DAVID SPIEGELHALTER

The Norm Chronicles (with Michael Blastland)

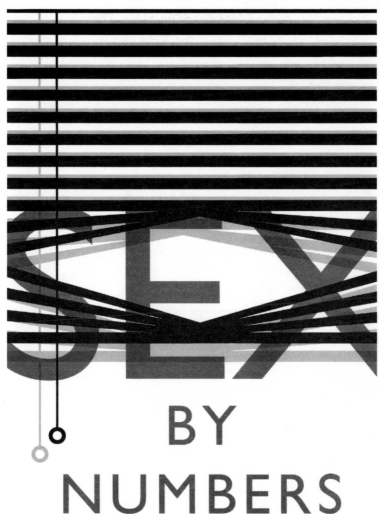

BY
NUMBERS

The Statistics of Sexual Behaviour

DAVID SPIEGELHALTER

P
PROFILE BOOKS

wellcome
collection

First published in Great Britain in 2015 by
PROFILE BOOKS LTD
3 Holford Yard
Bevin Way
London WC1X 9HD
www.profilebooks.com

Published in association with Wellcome Collection

Wellcome Collection
183 Euston Road
London NW1 2BE
www.wellcomecollection.org

1 3 5 7 9 10 8 6 4 2

Typeset in Palatino by MacGuru Ltd
info@macguru.org.uk

Printed and bound in Great Britain by
Clays Ltd, St Ives plc

A CIP catalogue record for this book is available from the British Library.

ISBN 978 1 78125 329 8
eISBN 978 1 78283 099 3

Mixed Sources
Product group from well-managed
forests and other controlled sources
www.fsc.org Cert no. TT-COC-002227
© 1996 Forest Stewardship Council

This book is dedicated to everyone in history who has struggled with sex. And eventually called it a draw.

CONTENTS

1

PUTTING SEX
INTO NUMBERS

Does oral sex count as 'having sex'?

Bill Clinton famously claimed on 26 January 1998 that 'I did not have sexual relations with that woman, Miss Lewinsky', a claim later repeated in a court deposition. It then became known that he had received oral sex from Monica Lewinsky. So did he or didn't he have sexual relations with her?

60%: the proportion of US students who thought that oral sex did not count as 'having sex'

What counts as 'having sex' might seem like a matter of individual opinion, but when Clinton was impeached for perjury in December 1998 – only the second time this had happened to a US President – it assumed national importance. In the same month the editor of the *Journal of the American Medical Association*, George Lundberg, fast-tracked a paper by researchers from the Kinsey Institute for Research in Sex, Gender and Reproduction Studies which was then published a month later in January 1999, just before the Senate impeachment hearing.[1] In 1991 over a thousand students had been randomly sampled from Indiana University, and

Figure 1: **What 599 US students thought of as 'having sex' in 1991**

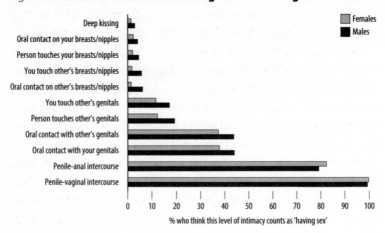

% who think this level of intimacy counts as 'having sex'

599 (58%) agreed to complete a history of their sexual activity and actually turned up to do so.[†]

As part of the sex history, the students were asked, 'Would you say you "had sex" with someone if the most intimate behaviour you engaged in was ...' – Figure 1 shows the responses. Just about everyone considered vaginal intercourse was 'sex' – the few men who answered 'no' are presumably waiting for some extraordinary activity before they feel they have gone all the way. At the other extreme, only a few considered that kissing breasts counted as sex. Around one in seven thought that 'sex' had occurred if genitals were touched, while 40% thought oral sex alone was 'sex', which means 60% thought it wasn't. So more than half would agree with President Clinton's claim of innocence.

Statisticians, contrary to popular opinion, are also human beings, and so I am fascinated by the special role that sex plays in our individual lives and society as a whole. Sex occupies a strange boundary between public and private: as President Clinton found out, sex can dominate news headlines

---

[†] That response rate may not sound great, but keep a mental note of it to compare with some other efforts.

yet (usually) goes on in private. We can speculate endlessly about the sex lives of others, but anyone trying to find out what is really going on will face a seriously challenging task.

But there are all sorts of reasons why we might want to know about sexual behaviour. It shapes the societies we live in: demographers, who study changes in population, want to know about sexual activity, and the use of contraception and abortion, so they can predict how many babies will be born and to whom. As we will see later, sexual activity may even shape the gender ratio of a population. Doctors and health researchers want to know what people get up to, and what precautions they take, in order to work out the chances of diseases being transmitted, and to plan the medical services for the unlucky ones. Psychologists may want to know about the quality of sexual activity and people's satisfaction with their lives. Psychiatrists want to identify and treat disorders, and pharmaceutical companies will want to develop and promote new treatments.

And the rest of us may be simply curious as to where we lie in the extraordinary range of human behaviour. Am I having too much? Not enough? With the right person? Did I start early, or late? Are my experiences different? Or at least, are they *really* different?[†]

Our sexual behaviour has a profound effect on how we live our lives: how society views you, whom you marry, whether you stay together, your health, whether you have children – all of these are shaped by sex. We are right to be curious. And we are right to wonder whether what we are told about sex – from government statistics to old wives' tales – is really what the numbers say.

---

† Of course, we have to face the prospect of finding out that everyone is having more sex than us. And that includes our partner.

How can we know what is going on behind closed doors?

To enjoy (or possibly suffer) any of the results of sex, you first have to have it. 'How much sex is going on?' seems like a simple enough question, but a moment's pause reveals that it is open to a variety of interpretations. We've already seen that people have widely varying ideas about what qualifies as 'sex'. We've left behind (although not that far behind) the time when sex between people of the same gender was not only socially stigmatised but actually illegal, so we can include same-sex sex. But what about solo sex? Whether or not you think that masturbation 'counts', later on we will count masturbation.

And when we are counting up sexual activities, do we include the (illegal) under-16s and the (legal) over-70s? And then there are different countries and cultures, and even the season can be important – we will see that Christmas holidays may be a particularly busy time.

So this simple question of 'how much sex' is already not so simple, and that's before we ask ourselves: how on earth are we going to find out?

A strictly scientific approach might install CCTV in a randomly selected set of bedrooms. This would not only make staggeringly dull viewing for most of the time but would also miss those sudden bursts of passion in the shower or the shed. So maybe we could put head-cams on some willing volunteers? Unfortunately, anyone who signed up to this experiment is hardly likely to be a representative sample of the population, and I doubt whether the study would get through a research ethics committee (although we are going to meet some very bizarre studies that presumably someone approved). And even if it did, this monitoring might encourage unusual performance, whether hesitancy or exhibitionism – the so-called 'Hawthorne' effect, when just scrutinising an activity changes what is done. Just think of *Big Brother*.

There are other, more reliable methods, though none of these is perfect. Whatever the sexual activity, someone, somewhere has tried to count it, but a running theme throughout the book will be the doubtful quality of many of the numbers that have gained headlines in the past: there's a lot of shabby statistics out there that keep on getting recycled. So in an attempt to provide some degree of order, I shall often give numbers a 'star rating' that says how reliable I think they are. Let's start in the top drawer.

4*: numbers that we can believe

We can get concrete evidence of some of the consequences of sex by counting babies, or treatments for diseases or other 'official statistics'. As it's a legal obligation to register a birth or marriage or abortion, these numbers should be reliable. So, for example, we can be confident that in England and Wales:

- 48% of births in 2012 were formally 'illegitimate'.
- In 1973, one in twenty 16-year-old girls got pregnant.
- For every 20 girls born, 21 boys are born.
- The peak rate for divorce is seven years after marriage.
- In 1938, half of brides under 20 were pregnant when they got married.

I shall label these as 4* numbers, which are so accurate that we can, to all intents and purposes, believe them. And we'll have a look at all these fine numbers later.

3*: numbers that are reasonably accurate

Nobody (yet) is under any compulsion to answer intrusive questions about their sex life, and so we are never going to be able to get 4* data about private activities. So we have to ask thousands of people about their behaviour and opinions, and try to do it well enough to be able to trust the answers.

It makes a big difference how the people are chosen. Suppose I want to know what proportion of people have sex before they are 16. I tell you that out of 1,000 young people, 300 say they did (this is about the current British estimate). If these 1,000 people had been chosen at random, with everyone in the population having an equal chance of being chosen, then a bit of statistical theory will show that we can be 95% confident the true underlying proportion of young people who had sex before 16 lies between 27% and 33%:[†] this relatively small margin of error is due to the play of chance in whom we happened to ask.

But if these 1,000 young people had been interviewed, say, coming out of clubs on a Saturday night, or had responded to an online survey in a lads' magazine, then I would have no idea what the error might be, except to suspect it might be large. Instead of pure random error, we have systematic bias. And it is this kind of bias that is so important in statistics about sex.

So try this little quiz. You have a sex life. Even if it's nothing to write home about, or so exotic that you would never write to anyone about it, you can still be a valuable data point for a researcher. If you were told the results would be confidential, would you feel happy answering questions about how often you had sex, what precisely you got up to, how many partners you had had, whether and how often you masturbated, and so on, if –

1. You were stopped in the street by someone from a market research company?
2. You were sent a questionnaire in the post?
3. A website for a magazine put up an online questionnaire, asking for volunteers?
4. You were part of an online consumer panel, and

---

† For the technically minded, this is based on $p +/- 2\sqrt{(p(1-p)/n)}$, where $p = 0.3$ and $n = 1000$.

this was one of the jobs offered to you for a small payment?
5. You were contacted by telephone by a market research company?
6. You were contacted by researchers wanting to interview you at home?

Would it make a difference if the survey were funded by a drug company or a condom manufacturer? Or if you were told it would contribute towards planning health services? And would it make any difference if you were paid, say, £15?

All these methods have been tried. But if none of these would incite you to participate, then you would be a missing data point. And if your reluctance to participate was in any way related to your sex life, then you would be biasing the results.

But there are good surveys, such as the British National Survey of Sexual Attitudes and Lifestyles (Natsal), properly conducted using random sampling and making repeated attempts to get information from individuals, using methods shown to maximise truthful reporting. And most of their results I would label as 3* numbers: reasonably reliable, with errors that are unlikely to make a substantial difference.[†]

For example, some Natsal statistics about Britain which we'll look at later include:

- The age at which the average woman first had sex dropped from 19 for those born around 1940 to 16 for those born around 1980.
- The average opposite-sex couple aged 16 to 44 had sex three times in the last four weeks.
- Around 70% of 25- to 34-year-olds had oral sex last year.

[†] I will rate a number as 3* if I judge that it is accurate to within a relative 25% up or down, so that a claimed proportion of 12% could actually be anywhere between 9% and 15%.

- One in five 16- to 24-year-old women has had a sexual experience with another woman.

2*: numbers that could be out by quite a long way

The next level of numbers tend to come from surveys that have not used random sampling, but where effort has been put into finding volunteers who cover a wide range of experience. Alfred Kinsey, perhaps the most famous sex researcher, obsessively collected 15,000 detailed sex histories in 1940s' America. Some of Kinsey's headline statistics that brought him notoriety, and which we'll meet in Chapter 4, included:

- 37% of men had had a homosexual experience resulting in orgasm.
- 50% of husbands had had extramarital sex.
- 50% of women were not virgins when they got married.
- 70% of men had had sex with prostitutes.
- 17% of men brought up on farms had had sexual contact to orgasm with an animal.

I would rate many of his results as 2*, which means they might be used as very rough ballpark figures, but the details are unreliable.†

1*: numbers that are unreliable

Even further down the scale come numbers that may be so biased as to be essentially useless as generalisable statistics, even if they do portray valid, and vivid, experiences. The

---

† Technical note: please feel happy to ignore all this. I shall take this as meaning that the true answer may be up to double, or as low as half, what is claimed. Proportions p should be changed to odds $p/(1-p)$, and the doubling and halving applied on the odds scale. For example a 2* proportion of 50% would be transformed to odds of $0.50/0.50 = 1$, doubled and halved to odds of 0.5 and 2, then transformed back to proportions of 0.33 and 0.66. So the true answer might be between 33% and 66%.

classic examples are the surveys carried out by Shere Hite, which were crucially important in the women's movement of the 1970s and 1980s. For her 1976 *Hite Report on Female Sexuality* she distributed 100,000 copies of her questionnaire to women's groups, chapters of the National Organization for Women, abortion rights groups, university women's centres and so on, followed up with advertisements for respondents in women's magazines.[2] She obtained 3,019 responses. This is a low response rate of 3% from a highly selected group, though to her credit Hite did not make much of the statistics, instead arguing from copious quotes that many women were dissatisfied with a mechanical male approach to sex, and that orgasm could be more easily achieved by masturbation than penetration. This report had a powerful influence on views of female sexuality in the 1970s.

Hite returned in 1978 with *The Hite Report on Male Sexuality* (7,239 responses out of 119,000 questionnaires),[3] and in 1987 with *Women and Love,* based on 100,000 questionnaires and 4,500 responses.[4] This time she heavily promoted her statistics, which included:

- 84% of women were emotionally unsatisfied with their relationships.
- 95% reported forms of 'emotional and psychological harassment' from their men.
- 70% of women married for more than five years were having affairs.

She received harsh criticism. *TIME* magazine put her on the cover but said the report was a 'male-bashing diatribe', while the Chairman of the Harvard Department of Statistics, Don Rubin, said 'So few people responded, it's not representative of any group, except the odd group that chose to respond.'[†][5] Unfortunately Hite continued to defend her

---

† Her statistics not only seemed rather implausible and out of line with other surveys, but were also just too neat to be true. Take, for

statistics as 'representative' when this was clearly not the case, and this provided a weapon for those who did not like her essential, and arguably very reasonable, conclusion: many women did not find their men communicative and loving, and thought they were too focused on the mechanics of sex. In any case, the statistical criticisms had limited impact: the lengthy personal stories (*Women and Love* runs to over 900 small-print pages) chimed with women's experiences and the books were best-sellers.

Although Hite's messages seem plausible, I would label her statistics as 1*: inaccurate.[†] Other 1* statistics that we will come across include the claim that single people in Los Angeles have sex 130 times a year, and that prostitution contributed £5.7 billion to the UK economy in 2012.

0*: numbers that have just been made up

We now get to the rock-bottom; numbers that get trotted out as part of an argument or to entertain, but have no supporting evidence. The sort of thing you might hear in the pub, on a radio phone-in or in Parliament. Some examples we shall deal with later include:

- Men think of sex every seven seconds.
- The average amount of time spent kissing in a lifetime is 20,160 minutes.

---

example, her conclusion that '70% of women married five years or more are having sex outside their marriages' – when broken down by ethnicity, the proportions quoted were White (70%), Black (71%), Hispanic (70%), Middle Eastern (69%), Asian American (70%), Other (70%). Such close agreement in proportions, particularly when some of the subgroups are very small, is essentially impossible.

† Technical note: I interpret 1* as meaning the true answer could well be more than double or half what is claimed, so a reported average of 4 sexual partners could in fact be greater than 8 or fewer than 2. An odds scale is used for proportions, so for a claimed proportion of 50%, the true answer could be greater than 66% or less than 33%.

- There are 25,000 trafficked 'sex slaves' in the UK.
- Expending an ounce of semen is the same as losing forty ounces of blood.

These I would rate as 0*: thought-provoking but utterly unreliable. And most misleading of all, of course, is the claim by Philip Larkin that sexual intercourse began in 1963 ('which was rather late for me'), but perhaps we should grant him some poetic licence.[6]

What can sex statistics tell us?

I am a statistician, and so this book will contain a fair amount of numbers and graphs (you might have been warned by the title). There will also be some extraordinary experiments, such as testing whether people admitted to having had more sexual partners when they believed they were wired to a lie-detector (women did), and whether sexual arousal reduced the disgust response (it did). But stats cannot say everything, so we will also hear stories from elderly people who were interviewed about their sex lives in the 1920s and '30s.[7] Such 'qualitative' data do more than bring colour to the statistics: they can remind us that every number is an inadequate attempt to summarise a collection of powerful and unique personal experiences.

1902: the date of the first published sex survey, on masturbation among members of the YMCA

Research on the statistics of sex has a colourful history. We will encounter William Acton counting prostitutes as he walked through London in the 1860s, F. S. Brockman producing the first published sex survey in 1902, on masturbation in the Young Men's Christian Association, and Magnus Hirschfeld asking Berlin metal workers about their sexual orientation in 1903. Then Kinsey shocking America with his extraordinary study, Shere Hite's analyses of female sexuality, and into the era of HIV/AIDS and the struggle to get serious surveys funded. These people were brave: sometimes risking physical danger and always risking public condemnation and their reputations.

These researchers thought of themselves as objective scientific investigators, but with hindsight were clearly driven by their own unquestioned assumptions and often a strong agenda regarding sexual politics. And so I should own up to being a white, middle-aged, middle-class man, full of implicit and explicit beliefs and assumptions, which are probably fairly representative of our current period and culture: a tolerance of alternative sexualities and behaviours between consenting partners, but a strong intolerance of sex involving coercion.

This book focuses on (fairly) normal behaviour. I have generally avoided what is currently illegal, including child abuse, incest and sex with animals (apart from Kinsey's data). Of course, some behaviour that has, in the past, been considered a product of a diseased mind will be included, such as same-sex activity and heterosexual anal sex. I also concentrate on sexual behaviour, so feelings and bodily responses get undeservedly small attention, although statistics on penis size and 'intravaginal ejaculation latency times' were too good to be left out. And although we will look at statistics about people's problems with sex, and also at some of the therapies suggested, you will need to look elsewhere for advice. I will also be sceptical of some attempts to answer more general questions about 'causes': what causes sexual orientation, whether exposure to pornography causes coercive sexual behaviour and so on.

I am happy to admit that statistics have their limits.

So did Clinton have sexual relations with Lewinsky?

55%: the proportion of the US Senate that agreed
that Clinton did not have sexual relations

Remember that in January 1999, as the US Senate prepared for the impeachment vote on President Clinton, a paper was published showing 40% of the students thought that oral sex was 'having sex', and 60% didn't. The press embargo on the paper was broken, and the media had a field day. Then when the Senate voted on the issue a month later, they split almost exactly the same way as the students had: 45 Senators said Clinton was guilty of perjury, and that he did really have sex with Lewinsky, and 55 said he was not guilty and that he was not lying when he said he did not have sex.[†] Clinton survived the vote and has become an elder statesman, while George Lundberg, the editor of the *Journal of the American Medical Association*, who had fast-tracked the paper for publication, was not so fortunate: he was sacked.[8]

All this discussion about 'What is sex?' may now seem a bit pedantic, but it has a serious point. When carrying out surveys of sexual behaviour, it is important to be clear about what is considered as a 'sexual partner'. If we want to know the risks of a sexually transmitted infection travelling through the community, we need to know how many people are at risk of catching something. So in modern surveys, 'having sex' has come to be defined as an activity that can transmit infections, and in Natsal an opposite-sex partner is someone 'with whom the respondent has had oral, anal, or vaginal sexual intercourse'. So, according to this modern definition, Clinton and Monica Lewinsky were sexual partners.

---

† Although it would stretch credulity to claim that this vote represented the true beliefs of the Senators – the vote was almost entirely on party lines.

Federal District Judge Susan Webber Wright agreed, and wrote 'his statements regarding whether he had ever engaged in sexual relations with Ms. Lewinsky likewise were intentionally false', and fined him $90,000. He later had his licence to practise law withdrawn for five years, and paid $850,000 to settle the sexual harassment suit that had started the whole business off. But did he 'have sex'? I'll leave it to you to decide.

2

COUNTING SEXUAL ACTIVITY

How much sex is going on?

On the face of it, this seems a simple enough question. But it gets to the root of the problem of researching sex, and it gives us a chance to see how people have struggled to get reliable answers.

23%: the proportion of British 25- to 34-year-olds who said they had not had sex with an opposite-sex partner in the last four weeks (Natsal-3, 3*).

A natural place to start is the most reputable recent British survey; the third round of the National Survey of Sexual Attitudes and Lifestyles (Natsal-3), which started collecting data in 2010.[†] Women respondents were asked 'On how

---

[†] A few details. The Natsal team sampled 59,412 addresses within 1,727 postcode sectors and sent preliminary letters and leaflets. Between September 2010 and August 2012 Natcen Social Research interviewers visited 26,274 eligible addresses and invited a randomly selected individual aged between 16 and 74 to participate – 15,162 completed the interview. The overall response rate was 58%, but 66% of those who could be contacted. Interviews were conducted

Figure 2: **Proportions reporting different frequencies of heterosexual sex over the previous four weeks**

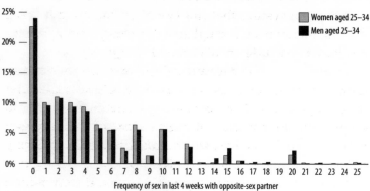

2,434 women and 1,500 men aged between 25 and 34 (12 responses with a frequency of more than 25 are not shown). Source: Natsal-3 in 2010.

many occasions in the last four weeks have you had sex with a man?', and vice versa for male respondents. And it was clearly explained that 'having sex' meant vaginal, oral or anal intercourse, so men who thought that kissing a breast was 'having sex' should have been put straight. In fact, the participants were not asked this in a conversation with a human interviewer – it was done on a computer.

Figure 2 shows the detailed data from Natsal-3 for women and men aged between 25 and 34 – the horizontal axis has been stopped at 25, since if I included one man's claim of '100 times' the whole picture would be crammed to the left.

There's a whole statistics lesson that could be based on this one graph. First, by comparing the heights of the bars, we can see that responses from women and men are broadly

------

using a mix of computer-assisted face-to-face interviews, and computer-assisted self-interviews. Interviews lasted around an hour, and participants received a £15 gift voucher. There was some over-sampling of 16- to 34-year-olds in order to increase the precision in that age group, and results were weighted by gender, age and region to match the national pattern. This survey has been carried out every ten years since 1990.

similar, which is reassuring. Then note the common use of round or almost round numbers such as 8, 10, 12, 15 and 20 times, suggesting that individual events are not being recalled, but an overall rough guess is being made, presumably based on a judgement of weekly frequency.

A particularly salient feature is the most common response, known as the 'mode': this is precisely zero. Nearly a quarter of 25- to 34-year-olds report having no heterosexual sex in the previous four weeks. Of course, some of these will not be heterosexually 'active', which Natsal define as not having had an opposite-sex partner in the whole of the preceding year: when we take out this group, we are left with around one in six 25-to-34s who are heterosexually active but have not had sex in the last four weeks – these will be primarily single people.

But while it's interesting to know what the largest number of people are up to, we might also want to look at the average person. Suppose we took all 1,500 male respondents aged between 25 and 34, and lined them up in order according to how many times they reported having had sex in the last four weeks. At one end would be a substantial crowd who reported no times, while at the other end would stand the man who claimed 100 times. The man standing half-way along is known as the median, who stands on 3, while the 'quartiles' are the man standing 25% of the way along (on 1) and the man at the 75% point (on 7). The same statistics hold for women, and so we can say that the 'average' 25- to 34-year-old reports having sex 3 times in the last four weeks.

Whereas the median tells us what the average person is doing, we can also look at what people are doing on average. This sounds confusing, but is simply a shift from the *median* to the *mean*: the total number of sexual acts, divided by the total number of people.[†] The 1,500 men reported a total of

---

† For example, if there were three men who had a frequency of 1, 3 and 11 times in the last four weeks, the median frequency would be

7,230 acts, so the mean is 4.8 per man. The mean for women is slightly lower at 4.4 – this is not necessarily an inconsistency, as some women will have partners older than 34, and some men have partners younger than 25.

Measuring an average by a mean can be misleading when there are some people reporting huge totals: if Bill Gates walks into a room (or even a small country), the mean income will change dramatically, but the median won't shift at all. Similarly a few claims of high activity, such as reports from sex workers, can have an undue effect – the individual who reported having sex 100 times in the last four weeks put the mean for 25- to 34-year-old men up from 4.7 to 4.8, all by his own efforts. So generally the median is the preferred measure of 'average'.

What influences the amount of sex we have?

2: the decline between 1990 and 2010 in the median reported frequency of sex in the last four weeks (Natsal-3*)

We can open this up to the whole age range questioned by Natsal, focusing on those who had an opposite-sex partner in the last year and so count as 'heterosexually active'. Figure 3 shows the summary statistics from the Natsal-3 survey.[1] Over the whole 16–74 age range, for both females and males, the median is 3 and the quartiles are 1 and 6, and so the average sexually active adult in Britain reports having sex 3 times a month, with the bottom quarter once or less, and the top quarter 6 or more times.

The obvious feature of Figure 3 is the steady decline of sexual activity with age, though this drop is not utterly

---

3 (the middle one), and the mean frequency would be the total, 15, divided by the number of men, 3: giving a mean of 5.

Figure 3: **Frequency of heterosexual sex reported in last four weeks by sexually active women and men**

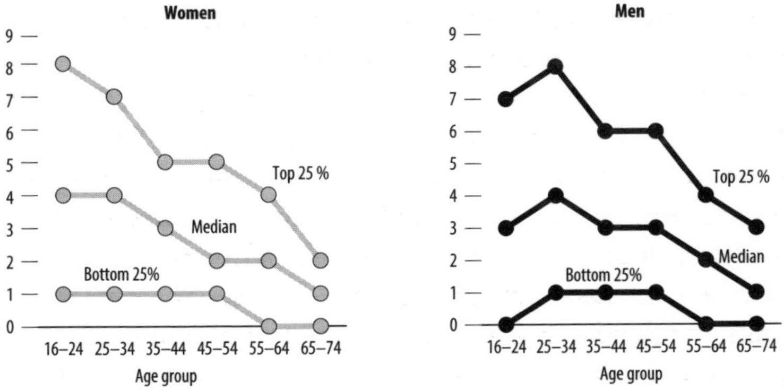

Source: Natsal-3 survey. The lines are the frequencies reported by people who were 25%, 50% and 75% along the spread of responses. So if you are a 40-year-old man who had sex 5 times with a woman in the last month, this puts you in the top half but not the top quarter of sexually active men of your age.

precipitous: although this may shock, and possibly even disgust, young people, being over 55 does not necessarily mean focusing only on gardening and trying to get your socks on without putting your back out. Older age can include considerable sexual activity: 64% of 55- to 64-year-old women reported a sexual partner in the last year, with a median frequency of twice a month and a quarter reporting at least 4 times: rather more men in this age group reported having a partner, 76%, with the same median frequency.

A question that springs to mind is whether the decline in frequency with age is because of ageing itself, or because older people will tend to have been in relationships longer, and maybe the initial energy and enthusiasm have worn off a tad. This certainly seems to happen for younger couples: 16- to 34-year-olds in Natsal-3 who had been in their relationship for less than two years reported a median frequency of 7 occasions in the previous four weeks, compared to a median of 4 occasions for those who had already been together for five years – that's nearly a halving of activity.

There are not many older people in new relationships, and

so it is not easy to work out how much of the decline above 35 is associated with being older, and how much with length of time people have been together. You might want to make your own judgement, but my feeling is that children and other responsibilities, and possibly sheer familiarity, can be as important as simply being older.

But as some of us are all too aware, being older often means your health deteriorates. So is this the cause of the decline in activity? Well, in part.[2] Natsal-3 showed that sexual activity declines with age, even for those in good health, but is lower for those in poor health, as was sexual satisfaction, regardless of how old they are. And, particularly sadly, although one in six people said their health had affected their sex lives, rising to two-thirds of those with bad or very bad health, only around a quarter of these affected people had tried to get help from a GP or specialist. We'll see more of this stoic attitude later.

The data from Natsal-3 provides a snapshot of what is happening in 2010, but invites the question – is the amount of sexual activity going up or down? Looking back at previous Natsal surveys gives the clear answer shown in Table 1; each decade has seen a steady reduction.

Table 1: **Median frequency of sex with an opposite-sex partner in previous four weeks**
16–44 age group who are heterosexually active

Year	Women	Men
1990	5	5
2000	4	4
2010	3	3

Source: Natsal-1, Natsal-2, Natsal-3.

At the rate of decline shown in Table 1, a simple, but extremely naïve, extrapolation would predict that by 2040 the average person will not be having any sex at all. I rather suspect this will not be the case, but this still leaves the crucial question: why is there less sex going on?

One possible reason for the decline in sexual activity is that fewer people are in partnerships and there are more people living on their own, without another body conveniently at hand. Having a live-in partner is certainly associated with more sexual activity: in Natsal-2 in 2000, cohabiting women between 16 and 44 reported a median frequency of sex of 6 times a month, married women said 4 times, divorced, separated and widowed said once a month, and the average single woman said she did not have sex at all in the last few weeks. But the frequency has also gone down among those with live-in partners – from a median of 5 in 2000 to 4 in 2010. This means that much of the decline shown in Table 1 is due, for some reason, simply to living in a more modern world.[†]

When the Natsal-3 results were announced in November 2013, one of the team casually mentioned that maybe people were just too busy with their fancy new tablet computers, and so the papers had headlines such as 'No sex, I'm on my iPad' and showed pictures of a couple sitting in bed, both staring at screens rather than at each other. In a recent TedX talk Catherine Mercer, one of the Natsal team, suggests that with 'increasing number of competing demands on our time, sex just falls down our list of priorities'.[3] Dealing with work emails late in the evening can mean that 'come bedtime, we're just still too connected with everyone and everything out there to be able to focus just on our partner'. It's impossible to say precisely why the decline has happened, but her suggestion that we are so busy and over-connected seems reasonable.

The pressures of modern life do not just involve electronics. Kaye Wellings, one of the leaders of the Natsal team, told me that middle-aged women were not having sex as they were 'totally knackered. Kids, parents who are sick, full-on

---

† It is challenging to separate out the effects of being older, length of relationship and the era in which we live, since all three increase together. We'll return to this conundrum when we come to anal sex in Chapter 4.

jobs.' Lack of interest in sex can be a simple consequence of an exhausting daily life.

How Natsal nearly didn't happen

~~~~~~

£7.3 million: the cost of Natsal-3 in 2010

~~~~~~

Natsal was almost stopped by attitudes to sexuality that would not seem alien to some Victorians. In the late 1980s there was massive concern about HIV: being infected by the virus was at that time seen as a death sentence, and its transmission crucially depended on sexual behaviour of the type that was generally disapproved of, such as same-sex activity, having many partners, casual sex without condoms and so on.

Nobody had much idea how much of that kind of behaviour was going on, and therefore how quickly HIV might spread. Some sort of survey was clearly essential, and the Natsal team was quickly created by merging groups with both medical and social science backgrounds. Despite some professional reluctance at being put together like a boy band, they swallowed their disciplinary pride,[4] and after a successful feasibility survey in 1988,[5] by February 1989 they were all set to roll their survey out. The team was in place and plans were laid: all they needed was the final government approval for funding.

And then there was silence. Nothing was heard for months. And then on 10 September 1989 came the *Sunday Times* front-page headline 'Thatcher halts survey on sex.' Mike Durham, the journalist who broke the story, 'had Margaret Thatcher, AIDS and sex all in one headline. If it was true, I couldn't believe my own luck.'[†6] Downing Street claimed the survey

---

†It seems that various ministers were unenthusiastic about the survey, but the Department of Health were very happy to shift

would be an invasion of privacy, unlikely to produce valid results, and would be an inappropriate use of public funds.

Fortunately, after lobbying from the Chief Medical Officer, Donald Acheson, the Wellcome Trust stepped in after a few weeks with £900,000 to rescue the survey. So, in spite of Margaret Thatcher, Natsal-1 went ahead in 1990, and this was followed by Natsal-2 in 2000 and Natsal-3 in 2010.[†]

How should you find out how often people have sex?

Getting the funding for your survey is difficult enough, but researchers also have to decide who to ask, what questions to ask, and how to ask them.[8] In a perfect world you could contact a large random sample of individuals, everyone would respond (or at least, there would be no bias in the sexual behaviour of the responders), the right questions would be asked in a way that elicits accurate answers and everyone would tell the truth. Not so easy. In fact, it's an unattainable ideal, so a huge amount of effort has been put into understanding the potential biases and trying to make sex surveys more reliable.[9]

---

the blame to 10 Downing Street. And Margaret Thatcher, who was beginning to feel rather insecure in her premiership, was in turn pleased to be seen as the bastion of moral values and take the credit for the block.

[†] Natsal-1 in 1990 took a probability sample of 29,802 addresses with at least one resident aged between 16 and 59: after being unable to contact 12%, and 25% refusals, there was a 63% response rate, with 18,876 face-to-face interviews and a self-completion leaflet for those with sexual experience.[7] Natsal-2, in 2000, had a more restricted age group of 16–44, and used self-completion computer-assisted self-interview. Questions were basically the same as Natsal-1. There were 11,161 interviews, a 65% response rate. Men were under-represented in both, and the sample was re-weighted to match the overall population. Natsal-2 was largely funded by the Medical Research Council, so it did get public funding. Natsal-3 was jointly funded by the Medical Research Council and the Wellcome Trust, with additional grants from the Economic and Social Research Council and the Department of Health.

Let's look at the first question: who do you ask?

The idea of a random sample or 'probability sampling' is that everyone has the same chance of being asked: Natsal does their best but does not include institutions, prisons or military, which in particular means some single men will be missed out (although I still rate their conclusions as 3*). But this kind of random sampling, then followed by visits and personal interviews, is difficult and very expensive, and so researchers have used quicker and cheaper ways to find out how often people have sex, though inevitably the results start to go down the star-rating scale.

One option is to send out a mass of questionnaires and analyse whatever comes back. Shere Hite was not the first to think of this: in the US, Katherine Bement Davis published her *Factors in the Sex Life of Twenty-Two Hundred Women* in 1929,[10] and said that, of 971 married women, 81% reported having sex at least once a week, and 9% at least once a day. It was a groundbreaking study – shocking for the time – but, like Hite, this is 1* or at best 2* evidence: 20,000 participants were approached through women's organisations and college alumnae registers, but responses from only 2,200 were recorded, and these were strongly biased towards those with higher education.

Rather than sending out questionnaires, another approach is to go out and hunt for people like biological specimens. This was Alfred Kinsey's method: in his survey of US married men in the 1940s,[11] he found a strong age gradient of frequency of sex, with 21- to 25-year-olds reporting a median frequency of 10 times in four weeks, while 51- to 55-year-olds said 4 times. This is about twice the Natsal rates, which, even given the absence of iPads, seems rather high.

Alternatively, you can just advertise a survey and wait for responses to come in: this is the classic 'write-in' used by magazines such as *Cosmopolitan*, of which the modern version is the internet poll. For example, *Time Out*'s reader's survey[12] reports a median frequency of sex of 10 times a month for people in a relationship, 7 if living together, 5 if married and once if single: even allowing for the readership to be young, this is roughly double Natsal's rates. The sample size is big – 10,042 people – but *Time Out* acknowledge that 'the sample consists of those who chose to fill out our sex survey online.' And people who are not having much sex may choose not to.

But the survey method that is coming to dominate popular sex statistics is the internet panel. Market research firms used to go out and stop people in the street for each new survey, but now they establish fixed panels that get used again and again for small payments or points to be redeemed later. It's easy to volunteer for a panel such as YouGov, or you can sign up for Amazon's Mechanical Turk and soon be answering 'adult' questionnaires for around US 50c a time.

130: the average number of times people in Los Angeles have sex each year, according to the Trojan Sex Census (1*)

An example of this genre is the 'Trojan US Sex Census', carried out by the US condom manufacturer, which attracted great publicity in 2011 when it announced that single people have sex on average 130 times a year in the USA, while married people clocked in at 109, and Los Angeles was the sexiest city, at 130 times a year, while Philadelphia was down at 99. A little searching reveals this is based on ten-minute online questionnaires from a national panel of 1,000, plus an extra 200 people in each city.[13] These numbers might be true for those who chose to earn their few cents for filling up the form, but they represent nobody but themselves, and barely merit a 1*, in spite of their attendant publicity.

The Natsal team have tried to see if internet panels could replace the (very expensive) personal interviews, and experimented with three different market research panels, providing over 2,000 participants each in 2012. But they found that the results from the panels differed substantially both from each other and from the Natsal-3 data. It was also difficult to get enough people to respond, and so Natsal have concluded that web panels are not reliable enough.[14] That's why I give studies based on web panels 2* at best.

Natsal have made a great effort to obtain accurate answers from as representative a sample as possible: some of their and other researchers' work is outlined in an Appendix on p. 305. It's been found that a good introductory letter is vital, stressing the importance of the survey for medical research, and moderate payment is sufficient. Natsal also look at consistency with official, 4*, data. For example, the proportion who admit having an abortion – a sensitive question – in Natsal is lower than would be expected from national rates: the ratio of reported to expected abortions is around 85% in Natsal-1 and 2, but 67% in Natsal-3.[15] But the birth rate is around that expected, which suggests there may be some under-reporting of sensitive information.

My own feeling is that the respondents to Natsal and other good surveys may be slightly biased towards the more sexually active, particularly in older people, and that therefore frequency of sex, for example, is, if anything, slightly overestimated. But there are compensating factors: higher-risk people will be harder to contact, and respondents may also downplay their less socially desirable activities. My completely subjective judgement is that, while the results may be to some extent biased, I would still trust them to around a 10% to 20% relative margin, and that's why I consider them 3* data.

So how much sex is going on?

999 out of 1,000: the proportion of opposite-sex
activity that does not lead to a conception

So back to the original question: how many acts of hetero-sexual sex are there each year, week and day? Natsal-3 tells us the mean frequency of heterosexual intercourse for ages 16–74 is just over 4 a month – let's say, an average of 50 times a year for the 82% of men between 16 and 74 who report having been in a sexual relationship over the last year. There are around 22,500,000 million men in Britain in this age band, so that means 18,500,000 who say they are sexually active. At an average of 50 times a year, that's over 900,000,000 bouts of sex each year in Britain, not counting same-sex activities or things that you get up to on your own (a more detailed analysis taking in to account different activity in each age group comes to around a billion acts a year).

The 900,000,000 annual British heterosexual acts work out at about 2,500,000 each day, 100,000 an hour, 1,800 a minute (around 2,500 a minute during hours awake, as we are not counting solitary events when asleep).[†]

There are only around 770,000 children born in Britain each year, although taking account of abortions, still-births and miscarriages this means there are around 900,000 conceptions. So these conceptions were the sole fruits of around 900,000,000 'coital acts' (as they are sometimes known in the research world). So only 1 in every 1,000 acts of hetero-sex ends up in a conception, and sex is 'non-procreative' in 999 out of 1,000 occasions.

[†] An average ejaculation contains around 300 million sperm and fills 3 mls, half an official UK teaspoon, so this amount of activity works out at around 5 litres (just over a gallon) of semen a minute when added up over all of Britain. But I think this may be taking the numbers a bit far.

The simple conclusion must be that 99.9% of sex is just for fun, and even the 0.1% that ends up in a conception will, we might assume, be just as enjoyable.

3

SPIN YOUR PARTNER

How many opposite-sex partners have people had?

Take a moment, and count how many (opposite-sex) partners you have had in your life, using the standard Natsal definition of having had 'vaginal intercourse, oral or anal sex'.
Maybe there are none, or a small list you can easily recall
by name: or maybe you need to count on your fingers (and
possibly toes), or perhaps you may need to make a rough
estimate as the memory of faces and bodies blur into one of
those composite pictures.

14: mean number of lifetime opposite-sex partners reported by
men to Natsal-3. This is twice the number reported by women.

Figure 4 shows the responses for 35- to 44-year-olds in
Natsal-3 in 2010, representing people born around 1970. The
most common response is 'one' – about one in six people
in this age-range have only had sex with one person. The
median response (the one half-way along the distribution)
was 8 for men and 5 for women. Note the spikes in the graph:
there is a strong tendency to use round numbers for counts
greater than ten, with peaks on 20, 25 and so on, suggesting

Figure 4: **Proportion of 35–44-year-olds reporting different numbers of opposite-sex partners in lifetime**

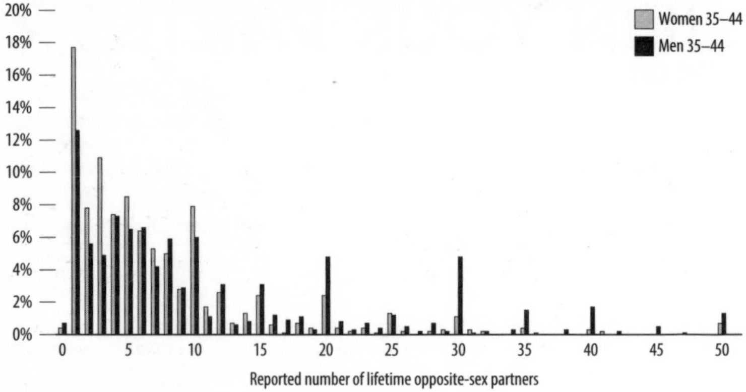

Reported number of lifetime opposite-sex partners

Source: Natsal-3 in 2010 based on around 2.000 interviews.

that a rough guess was being made rather than individual people remembered. The right tail of the distribution has been cut off: 6% of men and 1.4% of women reported more than 50 lifetime partners, and two men and one woman said 500, which was presumably a very rough guess indeed.

Figure 5 summarises these distributions for six different age groups, providing the medians and the 25% and 75% points – you can, if you wish, check where you lie. You might also want to consider whether you believe these numbers.[1]

Thirty per cent of women aged 25 to 34 reported more than 10 partners, compared with 8% of women aged 65 to 74, despite the older women having had more years to add to their collection, reflecting the changes in behaviour between women born around 1930 and those born around 1970. Having more than 10 partners in your lifetime is associated with higher social class and educational status – as we shall see in Chapter 7, the opposite effects from having sex earlier in your life. So women from higher socio-economic groups start their sexual activity later but then go on to have more partners.

You may have noticed the rather strange behaviour of the

Figure 5: **Summary statistics for lifetime number of opposite-sex partners**

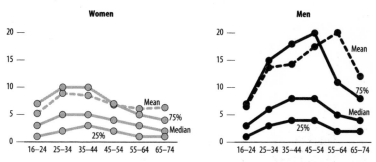

Source: Natsal-3, 2010. If, for example, you are a woman aged 30 who has had 11 partners, you are above the 75% point, and so in the highest quarter of your age group. If you have had 2 partners, you are 25% of the way up the distribution.

'mean' number of partners in Figure 5: it lies in the upper 'tail' of the distribution, especially for older people. This is not mathematically impossible: it simply points to the undue influence of a few people reporting vast numbers of partners and hence skewing the statistics, like including Warren Beatty or Jack Nicholson in the survey. But there is something about the mean statistics that is one of the great mysteries of sex surveys, and may shake your confidence in some of the stats we've seen so far.

Why do men report more partners than women?

In all sex surveys men report having had more opposite-sex partners than do women. For example, in Natsal-3 men reported an average (mean) of 14 partners over their lifetime, double the average of 7 for women. The 2010 Health Survey for England gave an average of 9 for men, again roughly double the average of 5 for women.[2] You might think this is a consequence of men getting around a bit more, but there is no avoiding a simple mathematical fact: in a closed group of equal numbers of men and women, the average number of partners over any period of time *must be equal for men and women*.

Figure 6: **Five men and five women, with 'partnerships' shown by lines**

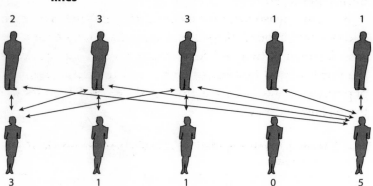

Also displayed are the total number of partners for each person. As predicted from the mathematics, the mean (average) number of partners is 2 for both men and women. But the median ('middle') number of partners is 2 for the men, and 1 for the women.

This may not be immediately obvious, but a picture should help. Figure 6 shows a group of equal numbers of five men and five women, with lines connecting those who have been partners – when teaching this to school students, these might be described as couples who have shaken hands, or danced together. Some have had no partners, while some have been shaking and dancing with relish. As each line connects one male and one female, the total number of lines represents the total number of partnerships – ten in this case. This is both the total number of male partnerships of the females *and* the total number of female partnerships of the males. So these two totals must be the same, and since there are equal numbers of men and women, the *average* number of partners must be the same, both for the men and the women.

This assumes that 'average' means the *mean*: the total number of partnerships divided by the total number of men (or women). In Figure 5 we have used a different sort of average: the *median*, which is the number of partners of the male (or female) who is half-way along the list, in the sense that half the males have had fewer partners, and half more. As was said when discussing the frequency of sex, the median represents the experience of the *average person*, and the mean the *average experience.*

It is quite reasonable for medians to differ between men and women: in our little picture the median number of partners for men is two and for women is one – in each case the third-highest number of partners. The median number of partners in the 16–44 age group is shown in Table 2 for the three Natsal surveys.

Table 2: **Median number of lifetime opposite-sex partners in 16–44 age group**

Year	Women	Men
1990	2	4
2000	4	6
2010	4	6

Source: Natsal-1, Natsal-2, Natsal-3.

There was a clear increase between Natsal-1 in 1990 and Natsal-2 in 2000 – some of this may reflect an increased willingness to report more partners, but most will probably be due to the relaxation following the 1980s' AIDS/HIV scare. There has been no recent increase.

There has been much soul-searching among sex surveyors about the suspicious anomaly in the mean numbers of partners.[3] The issues get to the heart of the difficulty of asking people these sensitive questions, and every possible explanation has been explored. The first concern is the people who get into the survey: there is a limited age range, and so some older men who have had a partnership with a younger woman may be left out, there is a gender imbalance in older groups due to earlier male deaths, and older men who may have fewer lifetime partners may tend to refuse to take part in the survey. All of these would mean men with fewer partners being left out.

Trying to adjust for these using some plausible assumptions does not fully explain the discrepancy, so maybe the answer is that many male partnerships have been with women who are not covered in the survey, either because they were abroad or because they were prostitutes. And including women working in prostitution would also push up the mean number of partners of females: as Kaye Wellings said to me, 'As far as we know, we did not get any prostitutes: one interviewer in the first survey did get to a brothel but they didn't answer.' A US study has claimed the discrepancy[4] can be completely explained by prostitution, but the Natsal team denies this.

We've mentioned that the mean is far more influenced by extreme values than the median, and some men do report very large numbers of partners: but although this will have some effect on the mean, getting rid of the extreme values does not have a big effect.[5] The discrepancy has also been attributed to the different way the sexes try to estimate larger number of lifetime partners: whether they recall and count specific individuals, work from a rough rate (say, 2 a year) or just give a general impression.[6] Males may just be vaguer, or perhaps they either consciously or unconsciously exaggerate their numbers, and women under-report, in order to fit in with social norms or to improve their self-image. The Natsal team showed that if women under- and men over-reported numbers of partners by about 4% per year, then this would explain the discrepancy.

But Kaye Wellings says her qualitative work showed little evidence of the stereotypical view of men rounding up, and women rounding down. Instead, she believes that by far the biggest explanation is in the interpretation of the question. In spite of trying to make sure the definition of a partner is clearly defined, men may broadly include events that would not be included by women, particularly if they were occasions they would rather forget, because of regret or coercion. Wellings also reports the case of a woman who only had oral sex in an affair, and did not count that person as a partner: 'She would not tell the interviewer something she would not tell her husband.'

All these factors may contribute to some extent, but my personal impression is that there is some tendency for men to over-report and women to under-report, mainly without a deliberate intention to mislead, both when using a 'counting' and when using a 'rough-guess' strategy for coming up with a number. The discrepancy is far smaller for short-term recall – the mean number of partners in the last year is 1.5 for men vs. 1.3 for women – but counts accumulated over a lifetime should be treated with some caution. But maybe there are some tricks to try to get people to be more accurate?

Can we trust people's replies?

30%: the increase in the number of sexual partners reported by US female students who thought they were attached to a lie-detector

People are not always accurate. Especially when they are asked about things they've done that they know others would disapprove of, and even more so if they suspect the data may not be confidential. This 'social desirability bias' may not be deliberate, conscious lying; we can't always bring all our experience to conscious thought, and when we do, we might recast with hindsight. Neither is it confined to questions about sexual behaviour. In 2008, people over 16 admitted to drinking an average of 12 units of alcohol a week in Britain,[†] although alcohol sales over the same period showed they on average were actually drinking over 20 units a week.[7] We can be confident that 40% of alcohol is not being poured down the sink, and some of this deficit may be poor memory or bad estimation, but undoubtedly much was what is known as 'social-desirability bias'.

[†] A unit contains 10 ml of pure alcohol, so a small glass of wine is about 1.5 units, and a pint of strong lager can contain 3 units.

When we answer sensitive questions, we may have in mind our image of ourself – what we might think of as a 'private' audience. In a slightly bizarre experiment, forty undergraduate women rated how enjoyable they found a piece of sexually liberal, but not overly erotic, fiction. But a random half of the group had first been asked to spend some time visualising and imagining talking to two student friends, while the other half were told to spend the time visualising two older members of their own family.[8]

Those who had previously visualised friends rated the fiction as considerably more enjoyable than those who had conjured images of grandparents, uncles and so on. This was only a small experiment but seems plausible – when you are next about to embark on a piece of sexual adventure, first imagine a favourite aunt (or even Margaret Thatcher) staring down at you, arms folded, and see if it makes any difference to your enjoyment.

So what could be done to stop people fitting in with the norms for their gender?

Well, if you thought you were wired to a lie-detector, you may be tempted to give more honest answers. Devious researchers took the usual gang of US psychology students trying to earn course credits and randomly allocated over 200 of them to one of three 'modes' while they filled in a questionnaire about their sexual behaviour.[9] A third were just left in a closed room in anonymity, and a third were led to believe that their questionnaire could be seen by another student supervising the experiment – this was called the 'exposure threat' mode – while a third were attached by electrodes to a completely fake lie-detector, complete with coloured pens drawing spurious wiggly lines on a roll of paper– these are known as 'bogus pipeline' experiments.

The results were what you might expect: women who thought they might be 'exposed' claimed an average of 2.6 sexual partners, the 'anonymous' group said 3.4, while those attached to the lie-detector said 4.4, at which point they roughly matched the male claim of 4.0. Men were also influenced by the 'mode', in a similar pattern but not as much as women, although the sample size was really too small to arrive at detailed conclusions.

Bookfuls of similar experiments have been done, which all point to the rather intuitive idea that we can only trust the answers if the respondent trusts the survey team to be respectful, confidential and serious with their data. Asking a room full of people 'Hands up who has paid for sex' is hardly likely to elicit an accurate response. Anne Johnson, one of the leaders of the Natsal team, feels the personal interaction with an interviewer is vital for the survey to be taken seriously – web panels just don't encourage this commitment in the same way. But respondents also need to be assured that the machine on which they have entered their private answers will be 'locked down' at the end of the interview and nobody will be able to see what they said. We may be willing to be a statistic, but few of us would like our intimate lives to become public.

How many people don't have sex?

There are many reasons not to have a sexual partner of the opposite sex. It may be because a same-sex partner is preferred. It may be a conscious choice, or it may be a product of circumstances. And it may be a source of regret or of relief: simple statistics are not going to tell us.

Figure 7 shows the Natsal-3 data on people without opposite-sex partners. More than half of women over 65 report not having had a partner within the last year, while for men in this age group the proportion is only 40%. These proportions drop to less than 10% for people in their 30s. The numbers are little changed since 2000.

Figure 7: **Proportions without an opposite-sex sexual partner**

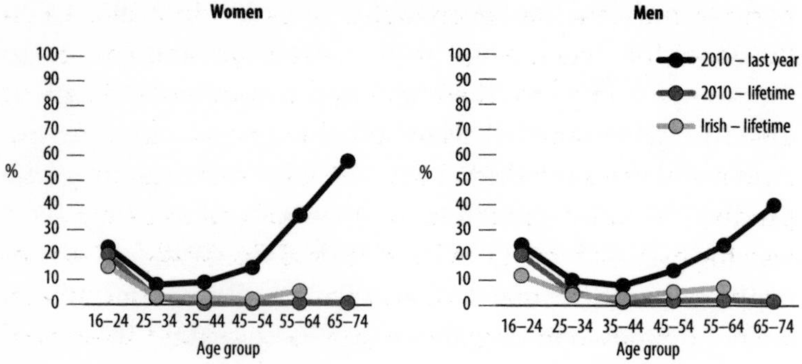

Proportions without an opposite-sex sexual partner in last year and over whole lifetime. Source: Natsal-3 in 2010 and Irish survey of 2004.

6%: the proportion of Irish 60-year-olds reporting that they have never had a sexual partner

The proportions who have *never* had an opposite-sex part-ner are also shown: these drop to around 2% of men and 1% of women by age 40, and some of these will have only had same-sex partners. This is almost certainly an underestimate of those in the population who have never had sex: Natsal does not include people in institutions, and I also suspect that older, celibate participants may be under-represented.

A somewhat different picture was found by the 2004 Irish Study of Sexual Health and Relationships,[†] whose lifetime data are also shown in Figure 7. They found that only 2% of 35- to 44-year-olds in Ireland had never had sex, just as in Britain, but this proportion then increased with older groups, so that for 55- to 64-year-olds, 7% of men and 5% of women had never had a sexual partner, and this is probably a considerable underestimate.

---

†This survey phoned 37,674 households in 2004, found 12,510 people within the 18–64 age range, and 7,441 completed computer-aided telephone interviews, for a response rate of 60%.

These were people who were born around 1945: previous periods in Ireland had even higher rates of adult celibacy. In the 1920s the average age at marriage in Ireland rose to 29 for women and 35 for men, and 24% of women and 29% of men had never married. Many became priests and nuns: the Irish researchers concluded that Catholic moral teaching had produced a state with the highest rate of non-marriage, and also the highest fertility, in Europe. Such high proportions of people who never married, and were presumably celibate, are not unknown in England: in the 1600s around 20–25% of the population never married, and although this dropped to around 10% in the 1700s, it rose again in the Victorian era.[10]

There are many reasons for not having sex. It's important to distinguish between those who are abstinent, possibly reluctantly owing to circumstances, those who are celibate from a positive choice, and those whose sexual identity is 'asexual': one of the less advertised findings from Natsal-2 is that 0.6% (1 in 160) of women reported that they 'never felt sexually attracted to anyone at all' – evidence of a small but increasingly visible asexual minority.[†] A lot of sex research is funded to find out about behaviour that could lead to pregnancy or transferring disease, which means that those who are not having sex remain a rather silent, but important, group.

How many people have two partners at the same time?

It's one thing to acknowledge that one has known, in the carnal sense, a moderate number of people, and quite another to own up to having juggled more than one over the same period. We shall see in Chapter 8 that people have become increasingly accepting about same-sex relationships and what consenting people do with each other, but at the same

---

† See, for example, the Asexual Visibility and Education Network (AVEN) website www.asexuality.org.

time have grown more and more intolerant of infidelity, or 'extra-dyadic' relationships, to give them their technical and slightly less judgemental name.

12%: the proportion of people aged between 16 and 44 who report having concurrent partners in the last year (Natsal-2, 3*)[11]

There have been a lot of extra-dyadics in literature and real life, from Anna Karenina to Prince Charles. But, being statisticians at heart, we don't just want stories – however entertaining – we want to know how much extramarital sex is actually going on. As you can imagine, this is not a straightforward statistic to obtain, and given the social disapproval that attends such behaviour, the way in which this question is asked can give rather different answers: words like 'affairs' or 'infidelity' or 'unfaithful' sound judgemental, and to get reasonably reliable answers it is recommended simply to ask whether you have had sex with someone outside your marriage or settled partnership.[12]

In the 1940s Alfred Kinsey did not beat about the bush; he just came straight out and asked how old people were when they first had sex with someone other than their husband or wife, and 50% of American males and 26% of females told him they had had extramarital sex. Shere Hite beat this with a claim that 50% of married women in the 1970s had affairs,[13] while *Cosmopolitan* in 1980 topped the lot with 69% of their married respondents over 35 saying they had had extra-dyadics.[14]

These figures do seem high. The US National Health and Social Life Survey (NHSLS) in 1992 reported much lower rates, 15% of married women and 25% of married men reporting ever having had extramarital sex, and less than 4% in the previous year. But not all of the NHSLS respondents were interviewed individually, and this can make a difference: only 5% of people interviewed with someone else in the room reported they had had more than one partner in the preceding year (whether extra-dyadic or not), compared with 17% who were interviewed on their own.[15]

Another recent survey of over 900 people found very similar results: 19% of women and 23% of men had 'cheated' during their current relationship, although these were volunteers answering an online questionnaire for the Kinsey Institute, so 2* data at best.[16] The researchers report that the men have 'an increased tendency to engage in regretful sexual behavior during negative affective states', while the women tend to have 'low relationship happiness and low compatibility in terms of sexual attitudes and values'. Another way to put this might be that the men tend to do something rash when they are fed up, while the women are just bored.

Natsal-2 found that 9% of women but 15% of men aged between 16 and 44 had had 'concurrent' partnerships in the last year; these statistics are medically important, as it should be obvious how concurrent partnerships can help transmit infections through a population. These events would not necessarily all be considered as infidelity by the respondents, as 'extra-dyadic' can cover a wealth of possibilities: people may have very different feelings about one-night stands, relationships that are only emotional – or only sexual – general philandering and so on.

Condemnation of sex outside a committed relationship is a social attitude, after all, and not everyone subscribes to it; these are very complex issues that are not captured by statistics. But out of all these studies, there does seem to be one reliable and consistent statistic: men are about half as likely again to have extra-dyadics than are women.

Who is the father?

One of the reasons that extra-dyadic relationships have traditionally been frowned upon is that they can cast doubt over paternity. The technical term for a simple mismatch between the assumed father and the actual biological father is the rather clinical-sounding Paternal Discrepancy (PD), but 'Is he really the father?' has been a standard topic of gossip, whether it concerns Mick Jagger, Leontes in Shakespeare's *The Winter's Tale* or Richard Wagner suspecting that his real father was Jewish. Of course, paternal discrepancy could be the result of generosity rather than subterfuge, through sperm donation, legal adoption or informally taking in the child of a relative or friend and treating them as your own, something that used to be common in Europe and is still practised in, for example, Maori culture.[17]

30%: the rate of Paternal Discrepancy found in paternity testing laboratories

In the past, suspicions might be based on the colour of hair or skin, or the shape of the body or the nose: now the results come from paternity testing laboratories, which will look at a set of genetic markers in the putative father and child and declare whether one could have come from the other. In around 30% of cases tested by these laboratories a paternal discrepancy is found, a conclusion that could have serious financial, let alone emotional, implications.[18]

But one of the worse abuses of statistics is when this is quoted as the proportion of *all* children that have a different father than is thought. This is an extreme example of selection bias: the group of children being tested is hardly a random sample of the population, as there was presumably some grounds for going to the lab in the first place. In contrast, a recent review concludes that, for the general population, the rate of Paternal Discrepancy is around 3.5%.[19] Which, of course, is still 1 in 30.

How long do partnerships last?

In spite of starting out with the best intentions, relationships can go wrong, and couples split up. Divorce used to be almost impossible to obtain, and in 1901 there were just 512 in all of England and Wales, though Figure 8 shows how much things have changed: in 2001 there were 141,135 divorces.[20] There was a big blip in the late 1940s after all those rushed wartime marriages, and then a massive increase in the early 1970s following the Divorce Reform Act of 1969 and the Matrimonial Causes Act of 1973, which meant that a couple just had to be separated for two years, instead of having to prove 'fault' – by being seen, for example, sharing a hotel room in Brighton with an extra-dyadic person.

***

42%: the proportion of British marriages
expected to end in divorce

***

Since 1985 the number of divorces has remained relatively stable and dropped recently. Meanwhile marriages have also been going down – it seems plausible that many couples who would previously have got married and subsequently divorced are now discovering their incompatibility when just living together.

Figure 8 shows that each year there are around half as many divorces as marriages. But this does not mean that 50% of all marriages are expected to end in divorce, even though this is sometimes claimed. It's a bit more complicated than that. In 2012 there were 118,000 divorces, out of around 11,000,000 marriages that existed at the time, a rate of around 1% of all marriages ending each year. That 1% is an average over all marriages, but it depends strongly on how long the marriage has lasted.

Figure 9 shows the proportion of all marriages that end in divorce for each year out from marriage. The highest failure

Figure 8: **Numbers of marriages and divorces in England and Wales since 1862**

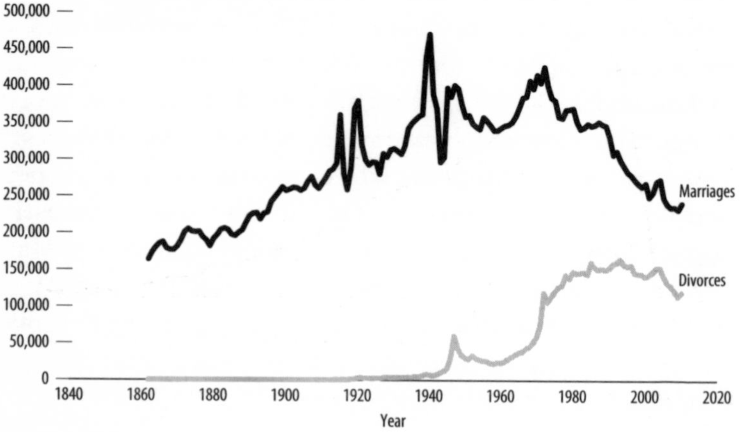

Source: Office for National Statistics.

rate for marriages is at around seven years: of marriages that have lasted six years, around 1 in 30 (3.2%) will fail in their seventh year. At last we have a mythical statistic that actually has a basis in reality: the 'seven-year itch', although presumably any itching started considerably before seven years. But if you can survive beyond seven years, then the annual risk steadily declines; this could be because the more robust relationships are left, a rediscovery of togetherness, or, more cynically, because some couples become too settled for it to seem worthwhile to split up.

So how long do marriages last? A popular error is to look at the median length of marriages among those getting divorced. This is eleven years. But this is yet another example of selection bias – it only looks at those that have failed, and ignores all those that are still carrying on. Using the divorce rates shown in Figure 9, the Office for National Statistics calculate that twenty years after marriage 60% will still be intact, 34% will have ended in divorce and in 6% someone will have died.[21] They also estimate that the average amount of time that a couple will expect to be together

Figure 9: **The proportion of marriages that will fail after each duration**

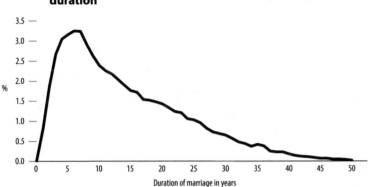

For example, of marriages that have lasted 10 years, around 2.5% (1 in 40) will fail in their 11th year.[22]
Source: Office for National Statistics.

before either death or divorce is a remarkable thirty-two years. Plenty of time to get used to each other's little ways.

The idea that marriage 'survival' has got worse over the last few decades is a much-repeated one, and it is true, but not dramatically so. Of those who got married in 1972, around 22% were divorced by 1987, fifteen years later. Jump forward twenty-five years, and we find that, of those who got married in 1997, nearly a third were divorced fifteen years later, in 2012, amid lessening stigma around divorce. But those that married in the last ten years seem to be faring a bit better. The risk of subsequent divorce is higher for those that marry young, those who were divorced before, and so on, and you can check your divorce risk using one of numerous online calculators based, for example, on data from the US National Survey of Family Growth (NSFG). If you say you are a single mother aged 20, who isn't Catholic and believes in divorce, you can hit the jackpot and get your ten-year divorce risk up to 91%.[23]

With current trends away from marriage we need to look at what happens to cohabitations too, which means using census data, as there is no official registration for living together. And marriages do turn out to be slightly more stable than cohabitations: of married couples in 1991, around 4 in 5 were still together in 2001, compared to around 3 in 5 of cohabiting couples (nearly half of whom had got married in the meantime): less stable cohabitations seem to be made by the young, those with no children, the less healthy, less well educated and unemployed.[24]

Civil partnerships for same-sex couples have been possible in the UK since the end of 2005, and there were 60,000 up to the end of 2012, at least five times the number envisaged in 2004. In 2012 there were 7,000, almost equally split between male partnerships (average age 40) and female (average age 38). This is much older than for opposite-sex marriages, and was even higher in the first year same-sex partnerships became possible, when the average age was 54 for men and 46 for women, as long-established relationships finally became formalised.[25]

But same-sex relationships have frailties too. There were 794 dissolutions in 2012 out of 53,000 partnerships established up to the end of 2011, about 1.5% in a year. Given that the average civil partnership at that time was of about 2 to 3 years duration, if we look at Figure 9 we can see that the failure rate seems very similar to that expected for marriages. However, unlike marriages, we have two types of civil partnerships, and those between two women seem to have about double the failure rate of those between two men. Although we might expect similar figures to be shown for same-sex marriages, which became legal in 2014, it is too early to tell.

I love statistics, but they do seem particularly inadequate at conveying the hopes, disappointments, trauma and possible recovery that are part of the formation and dissolution of relationships. But not every kind of pairing is part of a romantic attachment. Studies of US college students indicate that around half of those not in a long-term relationship have a connection that has been described as 'friends with bene-fits', also known as 'hooking up', in which there are physical encounters without commitment.[26] What happens to these dyadics? In a small study of 125 students, 36% stopped hav-ing sex but remained friends, 28% stayed friends with bene-fits, the relationship ended in 26% of cases, and 10% became romantic partners. So it was far more likely that the relation-ship would end altogether than it was to go on to become a committed romance.[27]

So these sexual relationships were not primarily tests before commitment – they were apparently for sex, friend-ship and, hopefully, fun. With the growth of apps such as Tinder, the idea of a sexual partner is becoming increasingly flexible. Statistics are struggling to keep up.

4

ACTIVITIES WITH THE OPPOSITE SEX

On an autumn morning in October 1950 three of the most eminent statisticians in the world sat waiting to meet Dr Alfred Kinsey in the Institute for Sex Research at the University of Indiana. To pass the time they continued to sing a jaunty song from Gilbert and Sullivan's *HMS Pinafore* they had started on the way to the Institute, but this impromptu performance was stopped abruptly when an irate Kinsey burst in and told them to be quiet. It was not a great start to the visit.

The statisticians were Fred Mosteller, who later founded Harvard University's Statistics Department, Bill Cochran, a 42-year-old Scot and a world expert in agricultural experimentation and surveys (and Gilbert and Sullivan), and John Tukey, a genius.[†] Kinsey had recently achieved international

---

[†] John Tukey was a prodigy and polymath. During the Kinsey period he was working with John von Neumann on early computers used in the development of the hydrogen bomb, during which he both invented the term 'bit' and organised folk-dancing at the uranium enrichment facility in Oak Ridge, Tennessee. In 1958 he became the first to use the term 'software' in print – he also developed the fast Fourier transform (FFT), invented the box- and stem-and-leaf plot and wrote countless academic papers. He could count in his

Figure 10: **A completed Kinsey form**[3]

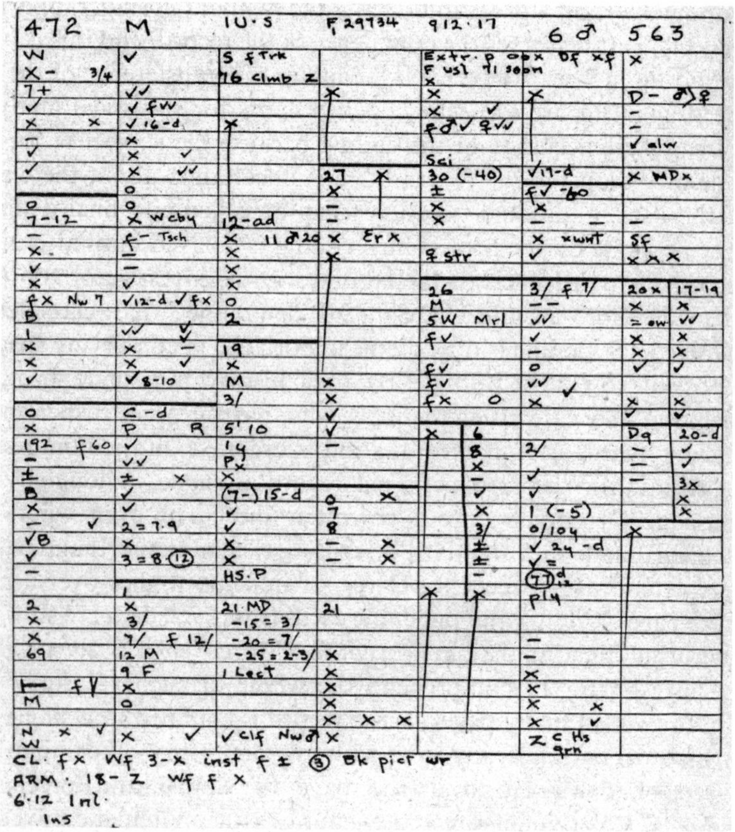

Answers to 500 questions were available at all times from a single sheet of paper – if you knew the code. Source: Kinsey Institute.

head and talk at the same time, a feat beyond even his student Richard Feynman: however, Feynman could simultaneously read and mentally count.[1] Tukey also solved the classic 'ham sandwich' problem: given a piece of ham and two slices of bread, each of any shape, is it possible to cut the sandwich with a straight knife so that the ham and both slices of bread are each divided into two exactly equal areas? He not only showed this was indeed possible, but also that you could do it to an *n*-dimensional sandwich, if you can imagine such a thing. Years ago I met Tukey at a dinner – a big man who 'ate pies for breakfast',[2] but he was very gracious to a young researcher.

fame, if not notoriety, for his study of the sexual behaviour of men, and the statisticians needed to understand how his data were collected. So they each had their sex histories taken using the Kinsey interview, an extraordinary piece of theatre.

As appropriate to the period, they were first offered a cigarette, and then rapidly questioned about every aspect of their sex life, their responses being unobtrusively noted on a 'recording sheet filled out in a most haphazard manner'. Figure 10 shows an example of the single sheet that allowed responses to up to 500 questions to be available at all times: the questions were not written anywhere, and the intricate coding included symbols whose meaning varied between questions: for example, 'lx(31)cx w Pr →✓ upset' meant that, when aged 31, the subject had extramarital sex with a prostitute and was very disturbed.[4]

15,970: the number of sex histories collected by
Alfred Kinsey and his colleagues by 1953

It was assumed that everyone had engaged in every type of sexual activity until it was explicitly denied: so a standard question would have been 'How old were you the first time you paid a female for intercourse or some other sexual activity?' John Tukey had only been married three months to Elizabeth, whom he had met while teaching folk-dancing, and so we can only imagine his response to 'During your marriage, how old were you the first time there was sexual intercourse with a woman other than your wife?'

The interview was never pruned of questions of doubtful relevance, so males were routinely asked 'Which testicle (ball) hangs lower?'[†] and which side of the trousers the scro-

---

† For the curious, of 3,305 white college-educated males, 21% said 'left', 67% 'right', 12% 'both equal'.[5] But did they know this without checking?

tum hung: as Kinsey's collaborators Gebhard and Johnson admitted in 1979, 'both of these matters would have been better dealt with by a urologist and a tailor'.

By October 1950 Kinsey and his team had carried out around 15,534 interviews – now it was 15,537, and somewhere in the Kinsey archive lurk our three statisticians' private histories. Their responses were punched on to cards by the interviewer himself, and Kinsey never allowed the coding scheme to be written down. After Kinsey's death it was realised that the code, and hence access to the data, was held in the heads of a few individuals 'all of whom could perish in a single plane crash or automobile wreck',[6] and so it was finally recorded and is now freely available online.[7]

Modern surveys have spurned his interview methods, but Alfred Kinsey can still be considered the founder of statistical sexology. His attitude is summed up by a famous quotation: 'We are the recorders and reporters of facts – not the judges of the behaviors we describe.'

Alfred Kinsey

Kinsey came from a very puritanical family and, when young, suffered intense fear of retribution if he masturbated. An obsessive collector, whether of irises, classical records or varieties of gall wasp, he collected thousands of specimens of the latter, recording their characteristics on a cryptic form. When he became Professor of Zoology at Indiana University, his long interest in the biology of sex led him to teach a 'marriage course' – essentially sex education – then increasingly common in US universities, although unthinkable in the UK at that time, or since. He found himself answering students' personal questions about sex, adapted his gall-wasp format into a questionnaire and in 1941 finally got some funding from the Rockefeller Foundation through the unambiguously named Committee for Research in the Problems of Sex (CRPS).

Figure 11: **Alfred Kinsey**

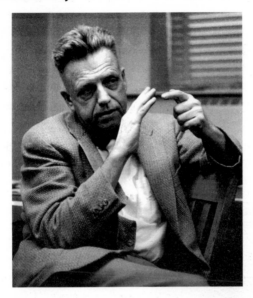

Source: Kinsey Institute.

He was obsessed with exploring every variety of behaviour, revelling in extreme cases and not considering anything abnormal or deviant: in particular, he regarded same-sex activity as simply a behaviour rather than as characterising a 'type' of person. He did not believe random sampling was feasible, and instead searched out groups of interesting subjects. He would, for example, go into the Salvation Army Home for Unmarried Mothers in Philadelphia and use 'a certain amount of pressure' to try to get 100% of them to co-operate. As Kinsey wrote, 'we go with them to dinner, to concerts, to night clubs, to the theatre … in poolrooms, in taverns and get them to introduce their friends' – he made a special effort to inveigle himself into the gay underworld, and collected many histories in prisons. He and his colleagues spent exhausting years travelling the country, finally collecting 18,000 histories, although he had really wanted 100,000.

Figure 12: **An illustration from *Sexual Behaviour in the Human Male***

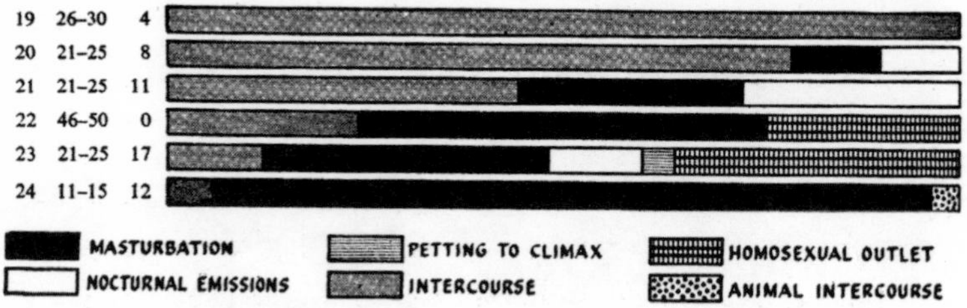

19	26–30	4
20	21–25	8
21	21–25	11
22	46–50	0
23	21–25	17
24	11–15	12

■ MASTURBATION   ▦ PETTING TO CLIMAX   ▨ HOMOSEXUAL OUTLET
□ NOCTURNAL EMISSIONS   ▩ INTERCOURSE   ▧ ANIMAL INTERCOURSE

An illustration from Kinsey, 1948 shows 'outlets' for six males at different ages and educational level: for example, for individual 24, who eventually had 12 years of education, between ages 11 and 15 his 'outlets' were primarily masturbation, but some with animals and intercourse. Source: Kinsey Institute.

The *Male* book,[8] based on 5,300 interviews, came out in 1948 and the *Female*,[9] with 5,940 interviews, in 1953. They are extraordinary documents, although rather verbose – each book is over 800 pages – with attractive line drawings and a huge number of tables. Each table could take a day to produce on a primitive card-sorter.

Details of Kinsey's findings are reported in other chapters of this book, but some headline claims that helped make him notorious were:

- 50% of American women and 90% of men had sex before marriage
- 26% of women and 50% of men had had extra marital sex, and
- 13% of women and 37% of men had had homosexual experience.

And that 1 in 12 men had had sex with an animal, and double that rate for farm boys. As mentioned in Chapter 1, I would, perhaps generously, consider that these are 2* numbers.

Both books are based on 'outlet' (orgasm) as the basic unit of analysis: Figure 12 shows the total outlet from six selected young men broken down into the proportions from different sources. These reveal the wide repertoire of reported activity in the US in the 1940s – we will see that the modern UK repertoire is just as varied, but somewhat different.

Vaginal sex

64%: the proportion of British women aged 45 to 54 who report having had vaginal sex in the last four weeks (Natsal-3, 3*)

Let's start, as most of us presumably did, with sex between a man and a woman,[†] also known as vaginal intercourse, coitus or a range of other less scientific terms. Kinsey concluded from his sample that between the ages of 16 and 40 more than 99% of married men have vaginal sex, which dropped to 83% between 60 and 65, and 70% between 66 and 70, with slightly lower rates in women. In Kinsey's day nearly everyone got married eventually, but he estimated that half of non-married men, and 35% of non-married women, had sex.

If we jump forward sixty years to Britain in 2010, Figure 13 shows the Natsal-3 data on those who had vaginal sex in the last four weeks, in the last year, or ever.[10] The rates are lower than Kinsey's, but perhaps more plausible. If we focus on the 'last year' curve, we see whole lifetimes summarised in a few inadequate figures. Women of around 20 are mainly single and most are sexually active, but then go on to form more settled relationships, largely cohabiting or marriage. By around 30 the proportion having heterosexual sex reaches its peak of 92%, where it stays for the next ten years or so. Then relationships start disintegrating, more women are single and there is a clear decline with age in the proportion who are sexually active: ill-health and widowhood then add

---

† Apologies to those who recognise this joke from *The Norm Chronicles*.

Figure 13: **Proportion reporting vaginal sex in last four weeks, last year and in lifetime**

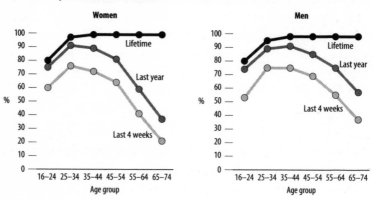

Source: Natsal-3. For example, of women aged between 35 and 44, 99% reported ever having had sex, 89% said it had happened in the last year, and 72% in the last four weeks.

to the decline, and so for women aged between 65 and 74, around one in three (37%) report having had sex in the last year, and around one in five (21%) in the last four weeks. This may be more, or less, than you expected.

This pattern is not so pronounced for men: the peak is slightly later, between 35 and 44, and the decline is not so steep, with men tending to find new sexual partners after separation or being widowed more easily. And, of course, these are not the same people that we are following along the curve: when today's 30-year-olds are 70, things may be different. We can but wait.

The UK survey only goes up to age 74, but the 2005 US National Social Life, Health and Aging Project (NSHAP) study included 75- to 85-year-olds.[†11] Using a broad definition of 'sexual activity' which does not necessarily include

---

† The US National Social Life, Health and Aging Project approached a probability sample of individuals aged between 57 to 85 in 2005, and interviewed 1,550 women and 1,455 men, an impressive response rate of 75%. Note the lack of the word 'sex' in the name of the survey, which can help prevent problems with funding.

intercourse or orgasm, they found 17% of women and 39% of men in this age group had been sexually active in the last year, around three-quarters of which usually involved vaginal sex – their figures for younger groups are similar to Natsal data, and so it may be reasonable to assume similar rates apply in Britain.

Sexual positions

Kinsey was, of course, interested in sexual positions. He pointed out that the standard practice recommended by authority – the missionary position – was not only very rare in other mammals but also was strongly culturally determined; it was not portrayed in the artefacts of early civilisations and, indeed, travellers from Western countries often found themselves as the object of amusement for local inhabitants (hence the name 'missionary'). He records that the 'woman on top' (cowgirl) position was that most commonly depicted in ancient Greece and Rome, and found it practised by about a third of college-educated US men in the 1940s,[12] although other innovations in pre-marital sex were rare. But sex in marriage got more varied, with around 27% of men and women trying 'side-by-side', facing each other, some or much of the time, and 12% doggy, or 'dorso-ventral' as it is referred to in Kinsey's tables.[13]

26%: the proportion of women answering the *Cosmopolitan* survey in 1980 who said that 'women on top' was their favourite sexual position

Perhaps unsurprisingly, the popularity of sexual positions has not been a great priority for subsequent serious academic analysis, so the field has been left wide open for some fairly dubious studies. One *Cosmo* report was based on 106,000 responses from a cut-out run in *Cosmopolitan* magazine during 1980, and asked questions about women's experiences of sex, orgasm, masturbation and so on. They report that 61% of their respondents prefer the missionary position, with 26% saying woman on top and 8% doggy.[†] Fairly similar results were obtained in a 1990s survey of US college students,[14] who voted for their favourite position as follows:

	Women	Men
• Man on top	48%	25%
• Woman on top	33%	45%
• Doggy	15%	25%

These are 2* numbers at best, but they are good enough to show a clear difference of opinion between genders, with the majority of men preferring not to be on top, whereas women would prefer him there. Let's hope couples can negotiate themselves to a happy compromise.

Oral sex

~~~~~~

3%: the proportion of the recommended daily zinc intake contained in an average ejaculation

~~~~~~

Oral sex hit the headlines with Bill Clinton and Monica Lewinsky, but they didn't come up with the idea on their own. Thirty years previously Kinsey had asked how many people had performed fellatio (oral sex on a male) or cunnilingus (when on a female),[‡] although more vernacular alternatives would have been used in the interview.

[†] *Cosmo*'s sample is far larger than Hite's, but all these respondents still only represent the *Cosmo* readership who feel like cutting out a questionnaire of 79 questions, filling it in and posting it back. In spite of the 106,000 responses, these are still only 1* numbers.

[‡] 'Fellatio' rhyming with *ratio* rather than *patio*, 'cunnilingus' rhyming with … *Charles Mingus*.

In Kinsey's sample of college men, 35% had received premarital oral sex, but only 16% reported giving it. He attributed this difference to the use of prostitutes, and indeed his sample of married was more balanced, as 43% had received, and 45% given, oral sex. His sample of married women reported rather higher rates, with 52% giving and 58% receiving, and even given the considerable limitations of his data this was a remarkable statistic for 1940s US marriages, particularly as it was illegal.

Kinsey, unsurprisingly, believed that oral sex was an entirely natural activity that educated people had rediscovered, writing that 'once again, it is the upper [educational] level which first reverted, through a considerable sophistication, to behaviour which is biologically natural and basic'.[15] But the evidence, such as it is, seems to go against this claim.

First, oral-genital contact is very uncommon in the animal world, although it is practised by bonobo monkeys and fruit bats. It had no strong role in Greek and Roman sexual life – it's featured with some satyrs on vases, but it was considered a contemptible rather than erotic practice.[16] In a popular history of erotic illustrations the 'French' method, as it was known in Britain, does not feature until the late 1700s, and then sparingly until the 1920s,[17] which does not mean that nobody did it, but that it was not considered particularly erotic and was probably uncommon – perhaps not least because the typical state of physical hygiene would have made it a fairly unappetising prospect.

By the time of Kinsey things had changed: the 'genital kiss' had appeared in Theodoor Van de Velde's celebrated marriage manual from 1928[18] mainly for overcoming fear in women: it was tentatively suggested that it could also be given to a husband, although it was approaching 'that treacherous frontier between supreme beauty and base ugliness'. But Van de Velde's manual, although a best-seller, did not manage to eradicate completely a long-held taboo.

In a very moving study Simon Szreter and Kate Fisher asked a group of older people about their sex lives as young people in the 1920s and '30s.[19] Only 3 out of 80 people were in any way positive about oral sex, with the middle class less condemning than the working class, who were either unfamiliar with the concept or frankly repulsed by it:

'Gill': Well, I still don't know what people get up to with oral sex. Sounds revolting to me, it really does. We certainly never did that.

'Gerald': Oh yes, disgusting ... And very dangerous ... the mind boggles. It was referred to as unnatural practices. And, er, with good reason – they are too.

The respondents repeatedly emphasised the importance of hygiene and cleanliness.

'Norman': Oral sex – I don't know anything about that now. I mean, before I had oral sex with somebody I would most likely go and give my winkle a good scrub.

Szreter and Fisher argue that, since nobody then had any idea what anyone else did, someone like 'Catherine', born in 1912 and who enjoyed receiving oral sex, thought that she was unadventurous and everyone else was doing more exotic things – she did not realise how rare she was.

So do people in Britain still find it disgusting? Apparently not. Figure 14 reports the proportions reporting having

---

†Szreter and Fisher recruited 89 working- and middle-class respondents from an affluent area (Hertfordshire) and a mill town (Blackburn), all born between 1901 and 1931, with the majority contacted through non-residential daycare centres. Open-ended interviews took place around 2000 – the respondents were between 70 and 97, mainly around 80.

Figure 14: **Proportions of all Natsal respondents reporting giving or receiving oral sex in last year**

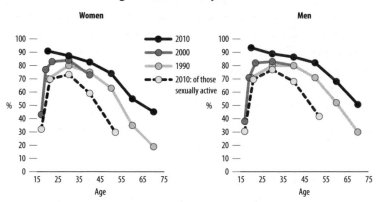

The dotted line shows the rates among those who were sexually active in the last year.

given or received oral sex in the last year in the Natsal surveys, who define oral sex as 'mouth on a partner's genital area'. There is a clear dependence on age, with around 80% of 25–34s saying they have given or received oral sex in the last year, compared with less than half of people around 60. Rates of oral sex went up both for men and women from 1990 to 2000, and then declined slightly from 2000 to 2010. But overall there's been a major change: in 1990 only 30% of women aged around 50 reported it, and this more than doubled by 2010: these were a new group of women who had grown up in the '70s. And as for Kinsey in the USA, Natsal found that oral sex is more popular among better-educated people and less popular in more deprived communities.

But do the rates of oral sex decline with age just because the sexually active population gets smaller? To answer this, we need to look at what happens to the proportion with oral sex as a practice, among those who are still sexually active. The dotted line shows this very well: it is clear that the popularity of oral sex steadily declines with age, and by around 50 only half of sexually active couples use oral sex as part of their repertoire. And some of the younger people may be only having oral sex, so that they may technically remain 'virgins' – we'll come to this in Chapter 7.

Oral sex got some bad press when Michael Douglas claimed it was responsible for his oral cancer.[20] But his chain of reasoning was not strong, to say the least, although in principle oral sex could raise the risk of getting this disease: the practice can transmit Human Papilloma Virus (HPV), and many oral cancers are associated with HPV infection.[†] But oral sex can also spread herpes, chlamydia, syphilis and gonorrhoea, but even if a particular behaviour may raise the risk of developing a disease, it is not possible to say what caused a specific case: we just don't know if they would have got the disease anyway by another route. And so Michael Douglas's assertion is a bit of an exaggeration, particularly in view of his years of drinking, smoking and drug abuse – all increasing the risk of cancer.

Finally, a popular sex statistic on websites is the number of calories in a typical ejaculation of 3 mls – numbers between 5 and 25 Kcal are quoted, but an estimate based on a detailed analysis of the chemical composition of semen[21] came to only 0.7 calories.[22] So if swallowed, it is non-fattening, and it does contain 3% of the US recommended daily amount of zinc.

Anal sex

16%: the proportion of British women aged 25 to 34 who report having had anal sex last year (Natsal-3, 3*)

Kinsey's reports caused a sensation. Hundreds of pages and tables and graphs documented the surprising extent of

---

† HPV lies behind genital warts, but there are many types of HPV, and high-risk ones can cause cancers, particularly cervical cancer; and since HPV is easily spread by sex, teenage girls in the UK, and also boys in the USA and Australia, are vaccinated.

pre-marital sex, use of prostitutes, extramarital affairs and bestiality. Conservative post-war America was shocked – or at least pretended to be shocked as it bought copies by the thousand. But there is one noticeable absence from Kinsey's list of 'outlets': there is no mention of anal sex.

How could Kinsey resist asking this? Of course, he couldn't – one of his questions was 'How often in the marriage was there anal coitus; putting the penis in the anus instead of the vagina?' Note the standard wording: not 'Have you ...' but 'How often ...' But he simply said in his reports that not enough data were available to estimate the incidence accurately.

While anal sex between men has been a recognised part of sexual tradition – though often with strong negative connotations – heterosexual anal sex has been seen as a rare aberration, although it has been represented in both Greek and Roman art, and Peruvian ceramics from around a thousand years ago.[23] But when people started asking, there was more about than was thought.

Kinsey's data, finally published in 1979, revealed that it had been attempted pre-maritally by 9% of his sample of white college men, and a remarkable (and rather implausible, given the proportion of their male peers who said they had practised it) 26% of his sample of white college women, while in marriage around 12% of college-educated couples said they had tried it.[24] Hite reported that 10% of her female respondents enjoyed anal sex,[25] while the *Cosmopolitan* survey said 13 to 15% practised it 'regularly'.[26] These figures were trumped by the *Playboy* survey of 1972, which claimed that 25% of married heterosexuals had had anal sex in the last year.[†27] But the 1992 US National Health and Social Life

---

† The *Playboy* survey was based on 2,026 adults gathered by a market research company which were claimed to be representative, but on which a certain amount of pressure had been used. A 'random' sample was asked to participate in a group discussion of sexual

Figure 15: **The proportion of all Natsal respondents reporting heterosexual anal sex in last year**

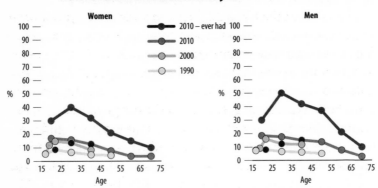

Also shown is the proportion in 2010 who reported ever having tried anal sex. The black dots represent the experience of a cohort of people born around 1970, who were aged 20 in 1990, 30 in 2000 and 40 in 2010. Source: Natsal-3.

Survey (NHSLS) rates were much lower: around 10% of people in their 30s had tried it in the last year, and 31% men and 24% women had ever tried it.[†] So by the 1990s there was

---

behaviour, and around 20% agreed. When they met, they were asked to fill out a long questionnaire and 'motivated by the discussion, and unwilling to walk out on the group, virtually 100% of the discussants completed useable questionnaires'. So even a 2* rating would be generous.

† The NHSLS was carried out in 1992, comprising personal 90-minute interviews, mainly with women in their 30s and 40s, with 3,432 respondents out of 4,369 eligible households, an amazingly high response rate of 79%. The resulting popular book *Sex in America*[28] boldly claimed to be the 'only comprehensive and methodologically sound survey of America's sexual practices and beliefs', and said 'surveys' such as Hite and *Cosmopolitan* 'are all so flawed and unreliable as to be useless'. The authors had a strong view that sexual behaviour was shaped by social surroundings, rather than being an innate biological function. Just as for Natsal two years earlier, the survey nearly did not happen. In September 1991 Senator Jesse Helms got an amendment passed that blocked government funding, but the team scrabbled together enough funds from charitable foundations. I would put the results from this survey as 2* or 3*.

conflicting evidence, with the poorer-quality surveys coming up with much higher rates.

Let's look at the UK data. Figure 15 shows the proportion of all Natsal respondents reporting heterosexual anal sex in the last year, and ever: the Natsal definition is 'A man's penis in a partner's anus (rectum or back passage).'[29] The graph tells a lot of stories about how behaviour can change with time, both as a person gets older and from generation to generation. Take 30-year-old men. In 1990 these men would have been born around 1960, and around 6% said they had had anal sex with a woman in the last year – slightly below the rates reported in the US NHSLS around the same time. The same age group in 2000 would have been born around 1970, and 12% reported anal sex in the last year. Thirty-year-olds in 2010 were born around 1980, and 18% reported anal sex in the last year. The figures for 30-year-old women are similar – a massive increase over a generation, although it is still the choice of a minority. The Natsal team also found that anal sex was practised less among more affluent people with higher academic qualifications – the complete opposite of oral sex.

The darkest lines show the proportion of people who have 'ever tried' anal sex, even just once – this was reported by around 50% for men and 40% of women aged 25 to 34 in 2010, and so born between 1976 and 1985. This is almost exactly the same as US data for 2006–10, where, for 30- to 34-year-olds, 48% of men and 37% of women said they had tried it at least once.[30]

Figure 15 shows clearly that the proportion who have 'ever tried' is more than double the proportion who said they had tried it in the last year, showing that, for many, anal sex is experimented with but does not become a regular part of their sexual repertoire.[†] If we compare the darkest lines

---

† This would make it one of many behaviours that are tried for the experience but do not necessarily become a habit. Like swimming at Blackpool.

in the two graphs, we see that more men than women say they have ever tried anal sex with an opposite-sex partner – this may seem paradoxical but is statistically possible if the women have had many sexual partners. Of course, there could also be some reluctance for females to 'admit' it if they felt some social stigma associated with the practice: this is 'social-desirability bias' again.

We can also look at what happens to a particular 'cohort' of individuals as they get older. The black dots indicate a group of people born around 1970, who were around 20 in 1990, 30 in 2000 and 40 in 2010. As this group has got older, their practice of anal sex has increased for men but levelled off for women. And a recent academic review of 'anal sexuality' also reported a range of other practices that Natsal would not label as full anal sex: for example, in a sample of over 1,000 heterosexual men who had not practised anal sex, around 10% had experience of insertion of a finger, sex toy or oral contact.[31]

The prevalence of anal sex in pornography has been cited as one reason for this change. But how can we ever determine why a group of people change what they do? It would be great if we could claim the statistics could show the way, but sadly when we try to find an explanation for any trend in sexual behaviour, we bang our heads against a basic logical problem.

What causes change in behaviour?

1980: if it is 2010, and you are 30 years old, you must be part of the 'cohort' who were born in 1980

We could imagine three competing theories that would explain differences in sexual behaviour. The first is that it is some biological imperative driven largely by gender and age – we can call this the 'age' theory. Second, it is driven by whatever the culture of the period is encouraging – call this the 'period' model. Finally, that it crucially depends on when you were born and where you grew up – we call this the 'cohort' explanation, since it says that your behaviour is determined by the group of people you come from.

So if were trying to pinpoint the broad factors determining, say, the proportion of people practising anal sex, we might look at their age (say 30), the period (say 2010) they are living in and the cohort of people born in 1980 that they were part of.

These three explanations – the 'age', 'period' and 'cohort' explanations – all have some plausibility. It would be naïve to think that just one or even two were in operation, and so it would be good to understand the relative importance of these three influences. Unfortunately, this turns out to be logically impossible – which presents rather a problem to researchers.

Let's suppose I enjoy something: say eating avocados. Is this habit driven by my age (60)? Or by the fact that avocados are currently popular and I am following the crowd? Or because people born in 1953 have a particular affinity to avocados? This last explanation is perhaps not as implausible as it sounds, as my post-war generation did not taste such exotic things until they were adults.

The problem is that it is logically impossible to separate out these explanations, since any two of them imply the third. For example, if I tell you when I was born, and how old I am, then I must be part of the current environment. Or if I tell you that I am alive now, and am 60, then it tells you when I was born.

It's clear that disentangling the 'reasons' for changes in behaviour, such as the increasing practice of anal sex, cannot be done by statistics alone. Although, as we shall see on numerous occasions in this book, this has not stopped people from trying.

BDSM

2%: the proportion of sexually active Australian
adults who engaged in Bondage-Discipline-Sadism-
Masochism (BDSM) in the preceding year

Although Kinsey had a personal interest in the connection
between sexual arousal and pain, he rather remarkably did
not explore this topic in detail in his surveys. But if he were
around today, he would doubtless be eagerly investigating
the world of BDSM. This can stand for Bondage-Discipline-
Sadism-Masochism, though the central DS can also stand for
Dominance-Submission. The term 'sado-masochism' will be
more familiar to many, but BDSM does not necessarily involve
the infliction of pain: it could involve a clear power relationship represented by, for example, one participant playing the
role of an animal and being harnessed and controlled. A Finn-
ish survey[32] listed the most popular practices as follows:

- bondage (89%)
- flagellation (83%) and
- handcuffs (75%)

and the most popular role-plays as

- master/madame–slave (56%)
- uniform scenes (39%) and
- teacher–student (29%).

In a major Australian survey, 2.2% of sexually active men
and 1.3% of women reported BDSM in the previous year.[†34]
In the spirit of Kinsey, the authors concluded that the behaviour was simply a lifestyle choice and was not related to

---

†The 2002 Australian survey[33] was of a representative sample of
19,307 respondents aged 16 to 59 using computer-assisted telephone
interviews, with a 73% response rate. I would rate it as 3*.

past abuse or difficulty with 'normal' sex. And a recent Dutch study of the personalities of 902 BDSM practitioners and 434 controls concluded that those practising BDSM 'were less neurotic, more extroverted, more open to new experiences, more conscientious, less rejection-sensitive, had higher subjective well-being, yet were less agreeable' than the controls.[35] The Dutch authors concluded that BDSM should be considered as part of 'recreational leisure, rather than the expression of psychopathological processes'.

You can now rent a plush dungeon in Hoxton containing a lavish array of equipment for £650 for three nights and two days, including a free bottle of Prosecco,[36] while online sex-aid retailer Lovehoney.co.uk recently sold 30,000 'restraints' in five months.[37] This has been a highly stigmatised practice, but there is a steady process of normalisation of BDSM into a recreational leisure activity. This could be termed 'the *Fifty Shades of Grey* effect'.

Other behaviours

41: number of different combinations of activities reported at most recent sexual episode (2*)

So far we have been considering an opposite-sex partner as someone with whom you share vaginal, oral or anal sex, but clearly all sorts of other things can be done with each other. These avoid not only the risk of pregnancy but also the possible transmission of disease, and the HIV/AIDS era renewed the interest in more 'hygienic behaviour': a 1980s' leaflet advertised '69 ideas to help you enjoy safer sex' including 'frottage' (body-rubbing) as number 1, 'washing each other' as 7, and 'fondling' as 28.[38]

People do not always have to have 'sex': Natsal-3 found that 'genital contact without intercourse' stays popular as people get older, with 41% of women and 56% of men aged between 55 and 64 engaging in this over the previous year. And people are varied and imaginative: the US National Survey of Sexual Health and Behavior (NSSHB) reported 41 combinations of sexual activity when describing recent sexual events, and although vaginal intercourse is still the most common behaviour, many sexual events included only partnered masturbation or oral sex.[†]

But this is still focused on the private parts, and ignores all the other signs of love and affection: kissing, hugging, cuddling and the rest of the 69 behaviours. There does not seem to be the same scientific interest in gathering statistics in these harmless activities, although *Glamour* magazine proclaimed: 'The average amount of time spent kissing for a person in a lifetime is 20,160 minutes. That's 336 hours, 14 days or 2 weeks. Lip balm anyone?'[39]

What a great statistic. But do we believe it, and where did they get it from? Let's say they assume an adult life of 55 years, then it works out as 366 minutes a year, which is 1 minute a day. So that's what they did – take an average of one minute a day and multiply it up. Do you spend a minute a day kissing?

A popular question is whether couples who display a lot of affectionate behaviour do better in the long term. Researchers from the Kinsey Institute in Indiana asked 200 couples in each of Brazil, Germany, Japan, Spain and the USA about how happy they were and their intimate behaviour with each other – the couples were aged around 50 and had been together for around 25 years, and the aim was to find what

---

[†] The National Survey of Sexual Health and Behavior (NSSHB) was sponsored by Trojan, the condom manufacturers, and carried out by the Kinsey Institute of Indiana. They used an internet panel set up by a market research organisation: in 2009, 11,000 adults and adolescents aged 14 to 94 were contacted, and 5,865 (53%) agreed to complete an online questionnaire. I would rate this 2*.

was associated with greater happiness.[40] Kissing, cuddling and touching were important – but for men far more than women. Sexual functioning was related to happiness for both sexes, but the cultural differences dominated everything: compared with the USA, Brazilian couples were far less happy, and the Japanese were happier. And, of course, they only got to interview the couples who were still together.

This is one of many efforts to find 'predictors' of satisfaction and so on. But they only found associations rather than causations. Rather than signs of affection leading to happiness, maybe it was the other way round: perhaps men were more touchy-feely because they were happier? Of course, this argument cannot always be used: people do not become Japanese because they are a happy couple.

Kinsey's fate

200,000: number of copies of Kinsey's *Sexual Behavior in the Human Male* sold in two months in 1948

Over 250,000 of Kinsey's 1948 male study were sold, and it was in the *New York Times* best-seller list for more than six months. But, just like Stephen Hawking's *A Brief History of Time*, it is doubtful whether many people read the book in any detail, and all the furore focused on what was in the early press reports: the men with experience with animals and so on.[†]

But a few people did read the details, and they did not like

---

†Kinsey finally reported that 8% of men and 4% of women had had sexual contact to orgasm with animals, rising to 40% to 50% for men living near farms having had sexual contact of some kind. The 1974 *Playboy* survey[41] claimed that 5% of men and 2% of women had had this experience. This change could arise from a reduction in opportunity because of fewer farms, or could be just unreliable statistics.

what they saw: statisticians and rival sex researchers claimed Kinsey's estimates were exaggerated and many of his conclusions were unfounded. Stung by the extensive criticism, in May 1950 the Committee for Research in the Problems of Sex asked the American Statistical Association to investigate, and so our merry team of statisticians ended up having their sex histories taken. Kinsey was intensely sensitive to any criticism but politely welcomed the investigating team, pointing out that whatever they said it was too late for the next book, on females.

In fact, Kinsey had already determined to take no notice of them anyway, and so any attempt directly to influence his work was a complete waste of time. But nevertheless the statisticians slogged on for years and eventually responded to Kinsey's verbose and over-detailed books by completing their own 338-page report written in a similar style, by which time it was 1954 and definitely too late to have any impact at all.[42]

The team identified numerous conclusions drawn by Kinsey that did not seem to be based on data. For example, take the claim that males have their peak 'outlets' in their late teens, and that female sexual desire and activity peaks in their 30s.[43] This has gone down in history as a 'fact', endlessly repeated, although his data seem to show no such thing: male activity is at an almost constant rate until 30, and there is no clear peak at all in women. In any case, we have seen that, although survey data show activity declines for both genders with age, this is partly due to familiarity in a relationship.

And, of course, the statisticians were particularly concerned about the lack of any attempt at taking a random sample: Tukey said, 'I would trade all your 18,000 case histories for 400 in a probability sample',[44] and a visit to the Kinsey family at home did not further endear him. Kinsey's wife, Clara, said, 'I never fed a group of men that I would have so liked to have poisoned [...] Tukey was the worst.'[45]

The titles of Kinsey's books – *Sexual Behaviour in the Human Male* and *Sexual Behaviour in the Human Female* – reveal the extent to which he believed he was objectively describing some innate aspects of men and women, and this 'biological' view has stimulated a new generation of criticism. Julia Ericksen's powerful feminist critique pointed out that he never considered that behaviour might have changed over the 20-year period of interviews, or that sexuality itself could be a product of social conditions:

> This man, whose wife gave up her career aspirations upon marriage, and who would not hire women interviewers because their husbands could not manage without them, did not understand that women's sexual responses were part of a larger ethic of service to men. Women could not afford an urgent commitment to sexuality, as this might interfere with what men desired.[46]

This chapter supports this view: major changes in sexual repertoire have occurred just over the last two decades, reinforcing the idea that sexual activity is so influenced by social context that it appears rash to claim it reflects some purely biological process.

But for many years Kinsey's data were all there was to go on when it came to discussing the statistics of sex. A more honest appraisal of their reliability could be made in 1979, when summary tables of a 'basic sample' of 11,246 out of 18,216 interviews were reported: around a third were removed as having come from prisons or from groups specifically recruited from the gay community.[47] This basic sample was 84% college-educated, since the working-class men originally included had almost all been in prison – Kinsey had not been concerned about this, as he thought that incarceration was typical of this class.

It is easy in the light of modern surveys to be dismissive

of Kinsey's efforts, but he was genuinely groundbreaking and completely changed the culture around sex research. But how many stars does his data really deserve? My own view matches that of my statistical heroes, who concluded that Kinsey's data were 'overextended in terms of scientific accuracy', but may have 'accuracy adequate for medical, legal and social decisions'. In other words, we can't trust the numbers to be precise, but the gist is probably broadly OK. I therefore accord the data 2*.

But by the time the statistical report came out in 1954, Kinsey's golden days were over. Critics staged an investigation into tax-free charitable foundations, but only targeted the Rockefeller Foundation. They caved in under this pressure and withdrew their funding. Kinsey, already suffering from poor health, struggled to try to raise money. He died in 1956, aged 62.

5

ACTIVITIES BETWEEN PEOPLE OF THE SAME SEX

Who was Magnus Hirschfeld?

As a gay, Jewish, left-wing, occasional transvestite – a term he invented – who openly campaigned for homosexual rights, Magnus Hirschfeld should not have been surprised that he was near the top of the list for Nazi persecution. Nevertheless, as he sat in a Paris cinema in May 1933, watching a newsreel of a mob wrecking his Berlin Institute of Sexual Research and then burning his library, he said it was as if he were 'watching his own funeral'.

Hirschfeld was born in 1868 into a prominent Jewish medical family, and qualified in medicine in 1892. In 1897, just two years after the notorious trial of Oscar Wilde, he started an unsuccessful campaign to repeal Paragraph 175 of the German penal code, which since 1871 had criminalised homosexual acts; his 'Scientific Humanitarian Committee' repeatedly managed to get it discussed in the Reichstag but ultimately failed in their attempts: it was not repealed in either West or East Germany until the late 1960s.

~~~~~~

2.2%: the proportion of males who are gay, as estimated by Magnus Hirschfeld in 1904

~~~~~~

In 1903 and 1904 Hirschfeld distributed over 7,500 post-cards to male students and industrial workers asking whether they were attracted only to females, only to males, to both, or were 'deviates' who did not fit into any of the three categories. Around half replied and, in spite of losing a court case for distributing obscene material, he eventually found that 1.5% of the students reported being homosexual and 4.5% said they were bisexual, with only slightly lower rates in the metal workers. And 6% reported themselves as deviates.[1] He then estimated the proportion who were homo-sexual in thirty-four different groups, including clergymen, English kings, actors and so on, and concluded that 525 out of 23,771 were homosexual – 2.2%; and this is the figure that he promoted.[2] It is worth remembering these figures to com-pare with later assessments using somewhat more scientific methods.

Hirschfeld started a sex journal in 1903, organised the first international conferences on sexuality, featuring speak-ers such as Bertrand Russell, and churned out a vast output of books and articles, becoming known as the 'Einstein of Sex'. In 1919 he opened the Institute of Sexual Research in the Tiergarten in Berlin, intended for science but which also became a clinic and a social centre for the thriving gay com-munity in Weimar Berlin. Before he wrote of the experiences that would be turned into the film *Cabaret*, Christopher Ish-erwood stayed at the Institute and affectionately described Hirschfeld as a 'silly solemn old professor with his doggy moustache, thick peering spectacles and clumsy German-Jewish boots', but regarded him as a 'heroic leader of his tribe'.[3] Hirschfeld needed to be heroic. This was a high-risk area in which to work, and he was repeatedly threatened and twice seriously attacked at his meetings.

He argued strongly that there was a biological basis for sexual orientation, and initially invented the concept of the 'third sex'. This term has stuck with him, although he moved

Figure 16: **Magnus Hirschfeld and Li Shiu Tong in Nice in 1934**

Source: The Hirschfeld Institute.

on to the more subtle idea of 'sexual intermediaries', an infinite variety of types between a 'full man' and 'full woman', including an important differentiation between dimensions concerning masculine/feminine *identity* and masculine/feminine *desire*. His *Yearbooks for Sexual Intermediaries* grew into a 20,000-page collection by 1923, featuring hermaphroditism, transvestism, homosexuality and so on: he was central to the growing European sexology movement.

On the personal side, his patient long-term lover was Karl Giese, but Figure 16 shows him with his eventual companion, Li Shiu Tong, trying to make the best of his final exile in France.

Counting same-sex activity

We must assume that same-sex activity has occurred in all cultures, with more or less discretion, and with often differing attitudes to the 'passive' and 'active' male. The ancient Greek and Roman relationships with younger men are well known, and the classic anthropological study *Patterns of Sexual Behaviour* reported that homosexuality was considered normal in 49 out of 76 (64%) of societies on which data were available, although attitudes could rapidly change when Christian missionaries arrived.[4]

2: the number of men (James Pratt and John Smith) publicly hanged outside Newgate Prison in London in 1835, convicted of committing buggery

---

But Western societies certainly did not consider the behaviour was 'normal'. Thomas Cromwell's Buggery Act of 1533 made sodomy, whether with a male or female, a civil rather than religious offence, with hanging as the punishment – the last execution was carried out three centuries later in 1835, after John Pratt and John Smith had been observed through a keyhole by a lodging-house landlord (and his wife).[†] However, although this specific act was condemned, it was not until the Victorian era that same-sex behaviour started to become a medical problem, a pathology, with the term 'invert' being used to describe someone trapped in a body of the wrong gender. In the late 1800s, the forensic psychiatrist Richard von Krafft-Ebing[‡] popularised the idea of a 'homo-

---

† As readers of Dan Brown's *The Da Vinci Code* may remember, in the 1100s the Cathars set up their strongholds in the south of France and practised their 'heretical' rejection of the bodily world, which included a ban on eating meat, Holy Communion, marriage and sexual reproduction. Part of the church's justification for the subsequent massacres of the Albigensian Crusade was the accusation that, since they did not believe in procreation, the Cathars must have indulged in 'unnatural' sexual acts, including sodomy. The Cathars were also known as Bulgars, as they were said to come from Bulgaria, and hence the term 'buggery'.

‡ Krafft-Ebing (1840–1902) believed any non-procreative sex was an

sexual' identity that was biologically determined in the womb: in the words of French philosopher Michel Foucault, 'the sodomite had been a temporary aberration; the homosexual was now a species'.[5] The British sexologist Havelock Ellis pleaded for tolerance, confidently identifying Erasmus, Michelangelo, Leonardo da Vinci, Christopher Marlowe and Francis Bacon as 'inverts'.[†] But the amount of same-sex behaviour, whether between men or between women, was a mystery, hardly surprising in view of its illegality (for men) and the social pressure to identify as heterosexual.

While it is an urban myth that lesbian acts were not made illegal in the UK because Queen Victoria did not believe they could exist, there was certainly less attention paid to female same-sex activity. Katherine Bement Davis's 1929 US study *Factors in the Sex Life of Twenty-Two Hundred Women* therefore created some surprise when it reported that 140 out of 1,000 single women (14%) and 234 out of 1,200 married women (20%) had experienced 'bodily exposure, mutual handling of organs, mutual masturbation or other intimate contact' with other women. As we saw on page 24, Davis's volunteer sample of US women was highly selected and hardly representative, but there was little else to go on until Kinsey came along.

---

unnatural practice, and eventually systematically analysed 238 case studies of 'perversions' or 'paraphilias' – he gave names to fetishism, sadism and masochism, which he thought were inborn conditions. He wrote details of the cases in Latin: the *British Medical Journal* could only 'regret that the whole had not been written in Latin, and thus veiled in the decent obscurity of a dead language'.

† Havelock Ellis was an empirical and pragmatic sex reformer, whose *Studies in the Psychology of Sex* tried to counter the mainland European obsession with sexual deviant pathologies, and who popularised a more tolerant individual and culturally relative view. He had essentially no sex life, and for a long time was aroused only by the sight or sound of women urinating, a condition for which he invented his own term – *undinism*.

The Kinsey scale

Compared with the simplistic idea of 'inversion', Hirschfeld's characterisation of a huge range of options of desire, identity and behaviour appears both subtle and humane. Kinsey was on Hirschfeld's side, and so constructed perhaps his most important lasting achievement: the Kinsey scale, shown below in its original version.

Based on both psychological reactions and overt experience, individuals rate as follows:

0. Exclusively heterosexual
1. Predominantly heterosexual, only incidentally homosexual
2. Predominantly heterosexual, but more than incidentally homosexual
3. Equally heterosexual and homosexual
4. Predominantly homosexual, but more than incidentally heterosexual
5. Predominantly homosexual, only incidentally heterosexual
6. Exclusively homosexual

Kinsey did include both experience and 'psychological reactions' in placing people on the scale, so it could be considered a composite of behaviour and attraction, although in modern use it often only represents behaviour.

The scale allowed a much more pliable description of sexual activity. Kinsey's headline finding was that 'at least 37% of the male population has some homosexual experience between the beginning of adolescence and old age', meaning physical contact to the point of orgasm. Also 10% of males are 'more or less exclusively homosexual for at least three years between the ages of 16 and 55' (i.e., scoring 5 or 6) and 4% of males are exclusively homosexual throughout their entire lives.[6] For 30-year-old US men, he estimated that 83% would score 0 (totally heterosexual), 8% would be 1 or 2 on the scale, and 9% would be at least a 3. He acknowledged that people could move on the scale during their lifetime, and indeed Kinsey himself is said to have moved from a 1 or 2 when younger to a 3 or 4 in middle age.

13%: the proportion of females who have had same-sex experience to orgasm, as reported by Kinsey (2*)

When he published his study on women, he estimated that 20% of women had had some same-sex experience and 13% to orgasm.[7] In unmarried females between the ages of 20 and 35, he claimed there was at least some homosexual experience (1 to 6 on the scale) in 11% to 20%, and 1% to 3% were exclusively homosexual. These results are remarkably similar to those of Davis in 1929, but both were largely based on educated volunteer samples.

Kinsey said he was totally unprepared for these findings, but claimed that he got similar answers whatever group was asked. And whatever the limitations of Kinsey's methods, these figures made big headlines, while decades later they would become political dynamite.

What proportion of the population is gay?

With the rise in gay liberation in the 1970s, the proportion of the population that was 'homosexual' became a hot political topic. Bruce Voeller, chair of the US National Gay Task Force, used Kinsey's data to claim that 10% of the US population were gay – this bold claim came from taking Kinsey's estimates for those with predominantly homosexual experience (4 to 6 on his scale) for at least three years, around 7% for women and 13% for men, taking an average and getting 10%.

10%: the proportion of the population that is gay, as claimed by the US National Gay Task Force in the 1980s

---

This was deeply controversial, and re-ignited old claims that Kinsey's data were unreliable as he had systematically recruited from prisons, gay bars and so on. But the numbers were reworked by Paul Gebhard, Kinsey's colleague, and in a letter to Voeller in March 1977 he stated that, even excluding these sources, 14% of male and 4% of women had had at least 5 partners or 21 overt homosexual experiences.[8]

The average of these figures is 9%, and so the National Gay Task Force continued with its 10% figure – as Voeller stated in 1990, 'The concept that 10% of the population is gay has become a generally accepted "fact" [...] As with so many pieces of knowledge (and myths), repeated telling made it so.'[†9] But Voeller also pointed out that the very flexibility of the Kinsey scale could backfire: many in the new gay movement began telling people who were 'only' 1 to 5 on the scale that they should make up their minds whether they were gay or not, being 'perceived at best as transitional to gay, at worst too cowardly to come all the way out of the closet'.[10]

This figure of 10% was strongly disputed by conservatives, who were therefore overjoyed when in 1991 the National Survey of Men came up with a very different figure for the proportion of gay men: 1%.[‡11] Statistics had become part of a political battleground.

The National Survey of Men had asked their respondents 'During the last 10 years, what would you say that your

---

† Bruce Voeller was also a biomedical researcher and author of some of the studies used in this book. He was at the meeting which decided the term 'AIDS' in 1982 – some recall he suggested the term – and died in 1994 of an AIDS-related illness in California, at the age of 59.
‡ The National Survey of Men was funded by the US National Institute of Child Health and Human Development, and took a probability sample of 4,700 men aged 20 to 39. Eighty-minute personal interviews with 3,321 respondents of all marital statuses were conducted in 1991, for an overall response rate of 70%. I would score it as 3*.

sexual activity has been?', with possible responses being: 1) exclusively heterosexual, 2) mostly heterosexual, 3) evenly heterosexual and homosexual, 4) mostly homosexual and 5) exclusively homosexual – rather similar to the Kinsey scale. 2.3% said that they had not been exclusively heterosexual, while 1.1% said they had been exclusively homosexual. In support of their conclusions, they quoted findings from the 1989 US General Social Survey that 'three percent have not been sexually active as adults, 91–93 percent have been exclusively heterosexual, 5–6 percent have been bisexual and less than 1 percent have been exclusively homosexual', which are very close to Hirschfeld's figures of 1.5% homosexual and 4.5% bisexual, although Hirschfeld was discussing attraction rather than activity.

Just because these estimates differ significantly, it does not mean that any of them are necessarily 'wrong' – because of the vitally important question of definition: what do we mean by 'homosexual' anyway? Do we mean identity (what sexual orientation you would claim), attraction (who you find sexually attractive) or behaviour (whether you have had same-sex experience with partners)? Much more subtlety was clearly required.

The 1992 US National Health and Social Life Survey (NHSLS) seems to have been the first to try and separate these aspects.[12] They concluded that 10% of men report at least one of same-sex identity, attraction or behaviour, but only 2.4% report all three – thus the figure of 10% might not be unreasonable, depending on the definition used. For women 8.6% report at least one, and 1.3% all three. And that often the behaviour has been transitory: 40% of men with same-sex experience had it before 18 and then it was not repeated.

It finally became clear that a huge amount of same-sex behaviour happens between people without a 'gay identity', which is why modern research refers, for example, to 'men who have sex with men' (MSM), rather than 'homosexual' or 'gay' men. These sorts of relationships are often associated with institutions such as schools, prisons and the military, though, of course, they are by no means limited to such environments. When discussing statistics, we therefore have to distinguish clearly between identity and behaviour. Let's start with identity.

Sexual identity

1.2 million: the estimated number of people in Britain who identify themselves as gay/lesbian or bisexual (3*)

Natsal-3 asked about sexual identity in the personal interview, rather than on the computer, showing the respondent a card and asking, 'Which of the options on this card best describes how you think of yourself? Please just tell me the letter next to the description on this card.' The options on the card were: (A) Heterosexual/Straight, (B) Gay/Lesbian, (C) Bisexual, (D) Other.

Figure 17 shows the responses[13] – the lines are not very smooth because the low numbers mean there is more variability due to the play of chance in who happened to be selected to be interviewed, so the fact that no women between 55 and 60 reported being bisexual is due to the small size of the sample rather than because there were really none out there. Combining the whole age range from 16 to 74, 1% of women and 1.5% of men said they were gay/lesbian, and 1.4% of women and 1% of men said they were bisexual. But there is clearly a gradient with age, with a higher proportion in younger people, particularly in younger women: 1 in 27 of women aged 16 to 24 reported that they think of themselves as bisexual or lesbian. It will be fascinating to see how this curve changes as the young women get older, and how they may differ from current 60-year-olds.

Figure 17: **Proportions identifying themselves as gay/lesbian or bisexual**

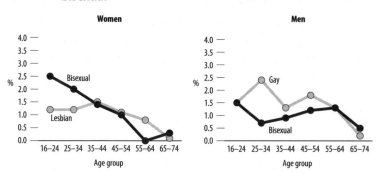

Source: Natsal-3 in 2010.

You may be surprised that sexual identity is now part of official government statistics. The Office for National Statistics (ONS) has recently introduced this question into their Integrated Household Survey, a huge effort bringing together six separate surveys with 180,000 respondents and a response rate of 63%, with face-to-face interviewers using computer-assisted personal interviewing.[14]

The ONS found 1.1% reported their identity as gay or lesbian – roughly the same as Natsal. Only 0.4% of the respondents reported themselves as bisexual to ONS, rather less than to Natsal, although the rates were much higher in younger people and in London. So is this evidence of bias in the Natsal participants or a sign of reluctance to open up in a general household survey that is largely concerned with mundane matters?

I am inclined to go with Natsal, and my judgement would be that roughly 1 in 80 adults under 75 would consider themselves gay/lesbian and 1 in 80 bisexual, but with the balance towards bisexual in women. For the UK population of 47 million between 16 and 75, this would mean around 350,000 self-defined gay and 230,000 bisexual men, and 240,000 lesbian and 340,000 bisexual women: a total of nearly 1.2 million – the population of Birmingham.

A recent US review reached similar conclusions: around 1.7% of US adults identify as lesbian, gay, 1.8% as bisexual, and an additional 0.3% as transgender, making around 9 million LGBT (Lesbian Gay Bisexual Transgender) Americans, roughly equivalent to the population of New Jersey.[15] But the routine US National Health Interview Survey (NHIS) of 35,000 adults recently came up with a lower figure for bisexual: 1.6% identified as gay or lesbian, and 0.7% identified as bisexual, while an additional 1.1% did not want to respond, said they didn't know or were 'something else'.[16] It seems plausible that, just as in the UK survey, there is less willingness to acknowledge bisexual identity in routine surveys, whereas those reporting gay or lesbian identity are willing to be more upfront with the interviewer.[17]

Same-sex behaviour

5%: the proportion of British women aged between 16 and 44 who report a same-sex partner in the past five years (Natsal-3, 3*)

When we move away from self-professed sexual identity and start asking about same-sex behaviour, it is crucial to be clear in our definitions: Natsal define 'same-sex experience' broadly as *any* sexual contact, which could be just kissing, while a 'same-sex partner' is someone with whom you have had 'genital' contact 'but not necessarily intercourse, intended to achieve orgasm, for example, stimulating by hand (mutual masturbation)'. Respondents are asked about activity at any age, including adolescence.

Figure 18 shows the how reported same-sex experience has changed between 1990 and 2010. For women in the age range 16 to 44, the proportion who report having had some same-sex experience has risen from 4% in 1990 to 10% in 2000 and to 16% in 2010 – apparently a massive change

Figure 18: **Same-sex behaviour reported in 1990, 2000 and 2010, ages 16–44**

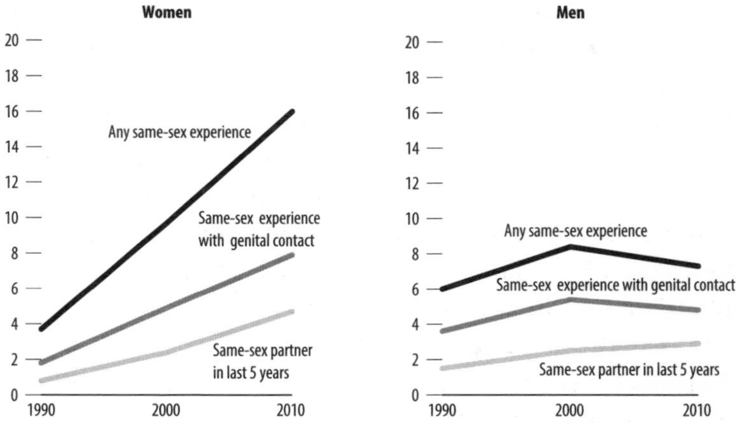

Source: Natsal-1, -2, -3.

in behaviour over such a short period. But this is not just girls kissing girls in imitation of Madonna and Katy Perry; around half report genital contact (thus fulfilling the definition of a same-sex partner), and around half of these in the last five years, so that overall nearly 1 in 20 females reports a same-sex partner in the last five years.

The question arises whether there has really been a change, or whether women might simply be more willing to report what they are doing. One neat check is to ask a question that should have the same answer in each survey. Natsal found, for example:

- the proportion of women reporting a same-sex experience before 1990, *when asked in 1990*: 4%
- the proportion of women reporting a same-sex experience before 1990, *when asked in 2000*: 7%

So the reported rate of same-sex experience before 1990 almost doubled between being asked in 1990 and being asked in 2000. Although these are not the same women being asked, and so the numbers would not be expected to correspond exactly, a rise from 4% to 7% is beyond the play of chance and strongly suggests an increased willingness to report sensitive information.[18] A similar analysis showed much more consistency between Natsal-2 in 2000 and Natsal-3 in 2010,[19] suggesting that the first part of the steep rise for women in Figure 18 was partly due to more honest reporting, but the later rise was real.

There is some increase in experience reported by men between 1990 and 2000, possibly associated with both better reporting and the decline in fear of HIV, but no substantial changes between 2000 and 2010.

There are even more marked differences between men and women when we consider the experiences of different age groups. Figure 19 focuses mainly on the proportion with same-sex partners over the last five years: it also includes 'ever had a same-sex experience' for 2010, although this will include transient adolescent activity. The proportion of women aged around 30 who have had a female sexual partner in the last five years has increased from around 1% (1 in 100) to 5% (1 in 20) between 1990 and 2010. There is a steep decline with age in terms of the absolute numbers with same-sex partners, but there has been a substantial relative increase in all age groups: back in 1990 the proportion of women around 50 reporting female partners in the last five years was negligible, but now it is around 1 in 40.

When it comes to their whole lifetime (the darker line in Figure 19), the proportion of women reporting any same-sex experience depends crucially on when they were born: nearly 1 in 5 women who are now around 20 report having had some same-sex experience, compared with 1 in 40 women who are now around 70. But it is clear that there is a lot of experimental activity – roughly, for each woman who has had a recent same-sex partner there are two more of the same age who have had some same-sex contact in their lives but no partner in the last five years.

Figure 19: **Proportion reporting same-sex partner in last five years, and in lifetime**

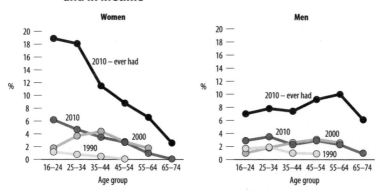

The darkest line is 'ever had same-sex experience'. Source: Natsal -1, -2, -3.

As we've seen, the proportion of men reporting recent same-sex partners rose between 1990 and 2000, but the behaviour of older and younger men is remarkably similar, with one rather interesting exception – there is a clear peak of lifetime experience for men around 60 and then a dramatic drop in those around 70, a pattern not observed in women. Working back in history might provide some explanation: those around 60 now were born around 1950 and were teenagers in the liberating 1960s, and in particular the legalisation of homosexuality in 1967: Natsal-1 reported a surge of first experiences in this group during this heady period.[20] Men who are around 70 now grew up when same-sex behaviour in men was illegal and frequently prosecuted, while much younger men would have come to adulthood in the more sober era of HIV.

But there is one characteristic that unites both female and male same-sex experience: the rates are two to three times higher among those with further education than those with no qualifications.[21]

89

How different are sexual identity and behaviour?

68%: of US women who have had same-sex experience, the proportion who nevertheless report themselves as 'heterosexual' (3*)

The US government survey that provides similar data to Natsal is the National Survey of Family Growth (NSFG), which has been carried out since the 1970s, its continued funding no doubt helped by its innocuous and worthy title.[†] Their published data provide a good opportunity to look at the complex relationship between same-sex attraction, behaviour and identity. And it will lead us to a rather startling conclusion: of people in the USA who have had same-sex experience, the *majority* identify themselves as 'heterosexual', and many say they are attracted only to the opposite sex.

Let's see how this apparent paradox can occur. Figure 20 shows the responses to relevant questions from the 22,600 respondents in the NSFG. A small proportion identify themselves as either gay or bisexual (3% men, 5.3% women), and a slightly higher proportion say they are attracted at least equally to the same sex. For both men and women, larger numbers report some same-sex behaviour. Note that the definition of a female sexual partner in the NSFG includes 'any sexual experience', whereas for men it only includes oral or

---

† Between 2006 and 2010 the latest NSFG interviewed 22,682 respondents aged between 15 and 44, with an impressive overall 77% response rate – 78% for women and 75% for men. Respondents in the 2006–10 survey were offered $40 as a 'token of appreciation' for their participation. Female interviewers visit homes but participants are asked sensitive questions through headphones and enter responses directly on to a computer – this audio assistance also solves language difficulties. I would rate this as 3* data.

Figure 20: **Reported same-sex behaviour, attraction and identity**

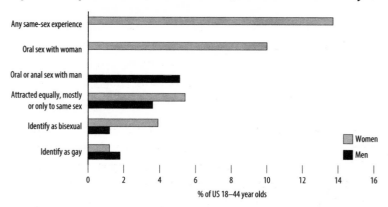

% of US 18–44 year olds

Source: the US National Survey of Family Growth 2006–10.

anal sex:[22] this is a somewhat odd asymmetry, but Figure 20 shows that three-quarters of women reporting such a partner also report having oral sex.

The crucial lesson from Figure 20 is that there must be many people who identify as heterosexual, or say they are only attracted to the opposite sex, who nevertheless have had same-sex experiences. Indeed, the NSFG report that of women aged between 18 and 44 who define themselves as heterosexual, 10% have had a same-sex contact (please note this number), and of those who say they are 'only attracted to the opposite sex', 5.5% report having had a same-sex contact. Of men who identify themselves as heterosexual, 3% have had same-sex contact, using the fairly stringent NSFG definition: of those who say they are only attracted to the opposite sex, 2.7% have had a same-sex contact.

So, overall, a small but notable proportion of self-professed 'straight' people have had same-sex contacts. We can now do some reasonably simple maths to prove our paradox.

Figure 21 shows what we would expect for 1,000 typical US women and men aged between 18 and 44. Consider women first. Figure 20 tells us that we would expect 5.3% to report themselves as being gay or bisexual, which comes

Figure 21: **A 'frequency tree' showing what we would expect from 1,000 US women and 1,000 US men aged between 18 and 44**

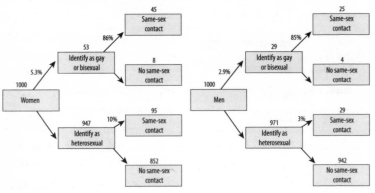

Source: US National Survey of Family Growth 2006–10.

to 53 women. Of these the great majority – around 45 (86%) – would have had a same-sex sexual contact. But let's look at the 947 remaining women who report themselves as 'heterosexual'. From the number I asked you to note previously, we would expect 10% of them to report a same-sex contact: a low proportion but this would still constitute 95 women. So the total number reporting a same-sex sexual contact is 45 + 95 = 140, of whom 95/140 = 68% reported themselves as 'heterosexual'. This is how the 'paradox' occurs: although same-sex contacts are fairly rare in those who identify themselves as 'straight', there are so many 'straight' women that these contacts still add up to a majority.

For men the figure is slightly lower, although this could be because the definition of a sexual contact is more stringent. We can calculate that, of men who report having had anal or oral sex with another man, a small majority report themselves as being 'heterosexual': from Figure 21 we can see this is 29/(25 + 29) = 54%.

We can also separately calculate that, of men who report having had anal or oral sex with another man, nearly half (44%) report themselves as being attracted only to women, while 32% of women who reported a same-sex contact said they were attracted only to men.

All this makes the historical argument over what proportion of the population is gay look rather naïve, although it's notable that the overall proportion for men and women who report same-sex contact in NSFG is nearly 10%, almost exactly Voeller's old claim from the 1970s. If we are anxious about sexually transmitted diseases, it's clear that the crucial issue is behaviour that can pass infections or viruses, and from this perspective whether someone actually identifies themselves as gay or not is almost irrelevant. Which is why the rather awkward phrase 'men who have sex with men' (MSM) has come about – the majority of men who have had sex with men label themselves as 'heterosexual'.

Same-sex behaviours

72%: proportion practising oral sex in a sample
of women who had sex with women

From a statistical perspective, general surveys of the type we've used so far will not be sufficiently accurate to estimate rates of specific behaviours, since the numbers with same-sex activity are too low. But it's very difficult to get a 'random sample' of gay men, although attempts have been made in US cities with a high gay population – San Francisco, New York, Los Angeles and Chicago – by having the survey steadily home in on areas with higher gay populations.[23]

Surveys generally concern volunteers who self-identify as gay or bisexual: for example, a huge online survey of nearly 25,000 men aged between 18 and 87, recruited from gay websites, reported that kissing on the mouth was the most common behaviour (75%), with oral sex and partnered masturbation coming up close behind at 73% and 68%. Anal intercourse was rarer, occurring among fewer than half of the respondents (37%).[24]

But how reliable are such volunteer internet surveys? The Natsal team tried comparing the group of men who have sex with men (MSM) found in their survey with those recruited via the websites gaydar and gay.com, using pop-up and banner advertisements on chatrooms and profile pages to ask for participants. The 2,065 MSM internet volunteers who were recruited were fairly similar to the Natsal survey in their age and backgrounds, but reported around four times the rates of sexually transmitted infections in the last year (16% vs. 4%) and increased rates of anal intercourse in the previous three months (77% vs. 63%). So it looks like internet recruitment can find volunteers who are representative in terms of their demographics but that may have higher-risk behaviour.

Less attention has been paid to same-sex behaviour in women. Bailey and colleagues offered 1,000 questionnaires about sexual behaviour to women attending lesbian sexual health clinics in London between 1992 and 1995 and got 803 responses, but got a rather lower response rate from lesbian community groups (415 out of 1,136).[25] The most popular practices were vaginal penetration with fingers (84% practised 'often'), oral sex (72%), mutual masturbation (71%), genital–genital contact (50%): other activities such as vaginal penetration with a sex toy (16%) were considerably rarer.

Natsal-2, in 2000, found a similar pattern in their small sample of 31 women who had sex exclusively with women, with 80% practising oral sex and 93% other genital contact in the past year. A larger group of 147 women who had sex with both men *and* women had somewhat different practices, with lower rates of both oral sex with women (54%) and other genital contact (61%).[26]

When we look at the increased subtlety with which

same-sex identity and behaviour are analysed, it's fair to say that there has been considerable progress since the simplistic idea of 'inverts' as a type of person; there has been increasing recognition of the extraordinary variety of sexual identities, attractions and behaviours, just as Hirschfeld recognised. But what determines your personal position in all this exhilarating diversity? Despite a century of argument and effort, this has remained a mystery.

What 'causes' sexual identity?

43: out of 53 gay people who were also an identical twin, the number whose co-twin was not gay (3*)

Although he was no scientist, Hirschfeld spent his life trying to 'prove' that sexual orientation was a product of biology. In contrast, Freud and his followers were firm believers that attraction to the same sex was acquired through upbringing and environment. One consequence of this latter view, of course, is that someone's sexual orientation might be 'changed' with appropriate therapy or a stupendous act of will. Indeed, William Masters, of Masters and Johnson fame, claimed to have converted 'dissatisfied homosexuals' to heterosexuality, although his colleagues later expressed scepticism about whether this actually happened.[27]

Soon after Kinsey's report on males was released, a classic study by F. J. Kallmann in 1952 appeared to give strong support to the biological theory.[28] Kallmann reported on 40 sets of identical twins, in all of which both members scored above 3 on the Kinsey scale (i.e., at least 'equally homosexual and heterosexual'), giving what is known as 100% 'concordance', whereas in 26 sets of non-identical twins there was only around 10% concordance. On this basis Kallmann claimed that sexual orientation was almost entirely genetic.

But his personal attitudes were clearly very different from Kinsey's acceptance of human variability, saying that further research was urgently needed 'as long as this aberrant type of behavior continues to be an inexhaustible source of unhappiness, discontentment, and a distorted sense of human values'.

Many similar studies of twins followed. For example, Whitam and colleagues collected 61 pairs of twins of whom at least one was not heterosexual. Identical twins were 71% concordant, non-identical twins were 40% concordant.

But these figures are deeply flawed. They may be only 1* numbers.

To understand this rather bold statement, we need to ask the question: why am I seeing a particular set of twins? The Whitam study advertised in the gay press for anyone who was a twin, and also got personal referrals from acquaintances.[29] But if both members of a twin-pair are gay, there is twice the chance of one of that pair seeing and answering an advertisement. And it is also reasonable to believe that acquaintances are far more likely to refer sets of twins who are both gay, and also twins with a shared orientation may feel happier volunteering.[30] So concordant pairs are likely to be grossly over-represented in the sample, and the overall proportion of concordant pairs is hugely biased through the selection process.[†]

[†] This is related to a well-known probability puzzle. I have two coins in my pocket: one fair, and the other two-headed. I take a coin out at random and, without looking at it, flip it. It comes down with a head showing. What is the chance that the other side is a head? The naïve answer is ½. But a little thought shows the correct answer is ²/₃. One way to get to this is to realise that the random choice of coin followed by the random flip gives rise to four equally likely possible outcomes, and in three of these a head will come up. But two of those heads will belong to the two-headed coin, and hence in two out of three situations in which a head shows after the flip, there will be a head on the other side. Essentially, when I observe a head, it's more likely that

Because of these biases, concordances were thought to be high until a landmark study of nearly 850 sets of identical twins in Australia was published in 2000.[31] They formed part of a national sample of twins, and their recruitment had been nothing to do with their sexual orientation. The researchers found that in 27 sets of identical male twins in which at least one was at least 2 on the Kinsey scale (defined as having at least 'substantial homosexual feelings'), only 3 sets were 'concordant': that's only 10%, far lower than previous studies. For 22 similar sets of identical female twins, again only 3 were concordant.

We can look at these results another way. Out of the 1,696 people who were identical twins, a total of 53 (3%) men and women identified themselves as having at least 'substantial homosexual feelings'. Out of these, 43 (81%) had an identical twin who did *not* identify themselves in this way. Simple selection bias seems to have led to forty years of grossly inflated claims about the genetic basis of sexual orientation.

The other problem with studying twins is that they generally will have shared a home environment, and so it's difficult to tease out whether any similarities are due to their genetics or their upbringing. That's why there's always been an obsessive interest in twins who have been reared apart, generally through being adopted into different families. Four sets of separated identical male twins with at least one non-heterosexual member have been studied: of these, two sets of twins were 'concordant'.[32] These are extraordinary stories. One man had gone to a gay bar in a neighbouring city, where he was mistaken for his twin, who he did not even know existed. The other set of concordant twins were also unaware of each other until they met by accident in … a gay bar. Imagine the scene.

These are Dickensian levels of coincidence, worthy of an

I have picked the two-headed coin. Similarly, when a gay twin replies to an advert, this in itself raises the chance they have a gay co-twin.

implausible drama. But from a statistical point of view it's important that the gay community is small, members are more likely to meet by accident, and hence the fact that they were both gay increased the chance that they would find each other. In other words, if only one had been gay, they would have been less likely to meet. Therefore the chance of concordant couples coming to our attention is again inflated, and there is more selection bias.

Overall, there does seem to be some genetic component to sexual orientation, particularly for men, but it is fairly weak. So if genes don't tell us very much, what about what happens after conception, while the foetus is developing? It's clearly an important time: it's possible to have identical twins where one has a normal heart and one has a serious congenital defect, owing to unknown influences in the womb. In a recent review Melissa Hines from Cambridge reports studies of females with 'congenital adrenal hyperplasia', who overproduce androgens (a class of hormones including testosterone): she concludes that the likelihood of their subsequently being bisexual or homosexual appears to be about 30%.[33] But Hines points out that there is no direct information on whether there is a relationship between *normal* variability in hormones during pregnancy and eventual sexual orientation.

But there may be a rough proxy measure. Stretch out your right hand palm-up – how long is your index (pointing) finger compared to your ring finger? In a massive online experiment run by the BBC, over 150,000 participants measured the length of their own fingers,[†] and on average their index finger was slightly shorter than their ring finger.[34] The ratio

---

† The precise instructions were: 'Hold your right hand in front of you. Look at where your ring finger joins the palm of your hand. Find the bottom crease. Go to the middle of this crease. Put the 0 of your ruler exactly on the middle of the bottom crease. Make sure the ruler runs straight up the middle of your finger. Measure to the tip of your finger (not your nail) in millimetres.'

between these two lengths is known as the 'digit ratio', and the BBC found that the mean digit ratio of men was about 1% smaller than that of women.[35]

It is thought that the relatively larger ring finger in men may be driven by testosterone exposure in the womb, and when we put together lots of studies that have measured fingers more carefully than in the BBC study, there is a clear difference between genders in their average digit ratios. But there is also a huge overlap: if a random man and woman are taken from a similar population, there is only around a two-thirds chance that the digit ratio of the man is less than that of the woman.[36] When we look within each gender separately, there have been numerous studies linking a lower digit ratio with more classically 'masculine' behaviours, so that men with lower ratios (shorter index fingers) tend to be more successful at financial trading,[37] but also have more traffic violations.[38]

When it comes to sexual orientation, analysis of many studies has shown that women identifying as lesbian or bisexual tend, on average, to have smaller digit ratios – that is relatively smaller index fingers – than those identifying as heterosexual: but with a large overlap. Although the BBC study found a weak association for men, the evidence is inconsistent.[39] So after all this work the conclusion is that pre-natal androgen exposure may have some influence on female sexual orientation, but little on men.

Another curious, but consistent, observation is that men with more older brothers have a higher chance of being gay or bisexual. For example, a study recruited 302 men from a gay community organisation and the 1994 Toronto Lesbian and Gay Pride Day parade and matched each of them to a man who was heterosexual and the same age, so that of the resulting 604 men, 50% were gay overall. But only 45% of those with no older brothers were gay, compared to 71% of those with four or more older brothers.[40] It's been estimated that each older brother is associated with an increased likelihood of being gay of around a third, so that three older brothers more than doubles the chance: a 'maternal immune hypothesis' has been suggested, where a mother's body 'remembers' previously carried sons.[41] These family associations are not found in females, who also, as in the data shown earlier, appear to have a more pliable and dynamic sexual orientation.

So where does all this mixed evidence on the 'causes' of sexual orientation leave us? The general consensus is now on the side of biology, but I don't envy those trying to disentangle the effects of genes, pre-natal hormonal exposure and early experiences – say, whether children conform to gender stereotypes.

It's a bit like spending a lot of effort trying to explain why some people are left-handed: up to the mid-1950s children were still being forced to change their 'orientation' and write with their right hand, but now it all seems of only academic interest. And ironically, there is an apparent association of handedness with sexual orientation, with the chance of not being right-handed increasing by around a third for gay and bisexual men, and almost doubling for women.[42] But then it gets even more complex, since the relationship between the number of older brothers and orientation has been claimed to hold only for right-handed men.[43] To be honest, it all seems a bit of a mess.

My personal view is that, while all this may be fascinating science, it is fruitless to search for a 'cause' in general, and particularly pointless for specific individuals. We can find associations, but the role of pure chance will always be so strong that all these statistics will have almost no predictive value for what your child's orientation will be. And that's just fine.

Hirschfeld's legacy

20,000: the rough number of books from Magnus Hirschfeld's Institute that were burned in Opera Square, Berlin, on 10 May 1933

Magnus Hirschfeld was tactless, impulsive and, to quote Kinsey's biographer, 'physically repellent', while Freud considered him 'flabby' and 'unappetising'.[44] Nevertheless he attracted loyal followers and was a popular speaker, although after 1930 the Nazi threat had grown to such an extent that he did not dare return to Germany after an international speaking tour.

He had good reason for his caution. He was an exile in France when he saw the newsreel of the events of 6 May 1933: his Institute was one of the first targets of a mob of Nazi students, who ransacked it as an emblem of 'degenerate culture' and a few days later carted the library and papers on to a huge bonfire in Opera Square. After the execution of the Brownshirts' leader Ernst Rohm in the Night of the Long Knives in June 1934, the persecution stepped up: if Hirschfeld had returned to Germany, he would have ended up either dead or in a concentration camp, wearing the pink triangle reserved for homosexuals.

He died in Nice in 1935, on his sixty-seventh birthday. Although he was no scientist, and his attempts to find a biological basis for sexuality failed, Hirschfeld's view of sexual identity and attraction as a flexible range of possibilities has stood up well. And in hindsight his statistical estimates look rather good, even if his survey methods were somewhat primitive. He was an extraordinary, brave and difficult man, and is increasingly feted in Germany and elsewhere as one of the founding icons of gay liberation.

6

BY YOUR OWN HAND

The great masturbation panic

57%: the proportion of young Christians in 1898
who admitted in a survey to having been tempted
by masturbation in their schooldays

Masturbation was invented – give or take a year or two – in 1712. Before that, people had presumably found that their private parts were at a convenient arm's length, but there was no great concern about what they then did with their discovery. Indeed Galen, the Greek physician, recommended masturbation for women to relieve bottled-up 'acrid humours', while Diogenes the Cynic was renowned for relieving his surplus bodily fluids in public.[1] The practice came under the general early Christian denial of non-procreative sex but was considered a fairly minor sin, more a subject of ribald humour than condemnation.

It's impossible to know how much went on except in occasional – very private – diaries. Samuel Pepys reveals repeated use of his 'fancy': on 15 July 1663 he wrote 'To bed, sporting in my fancy with the Queen', while the church was no barrier to

Figure 22: **The title page of 'Onania'**

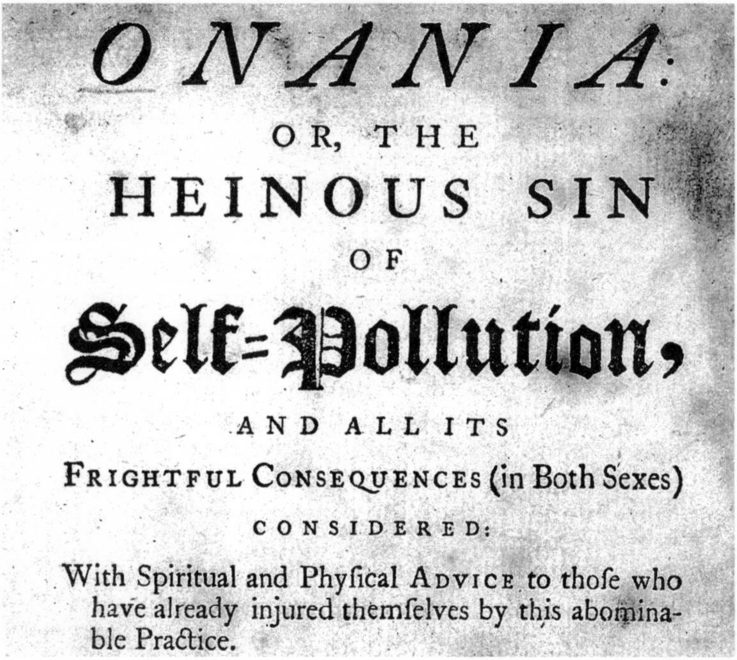

ONANIA:
OR, THE
HEINOUS SIN
OF
𝕾𝖊𝖑𝖋=𝕻𝖔𝖑𝖑𝖚𝖙𝖎𝖔𝖓,
AND ALL ITS
FRIGHTFUL CONSEQUENCES (in Both Sexes)
CONSIDERED:
With Spiritual and Phyſical ADVICE to thoſe who
have already injured themſelves by this abomina-
ble Practice.

Source: Google Books.

his hand, although he did feel the need to use some linguistic code: 11 November 1666 '… Anon to church, my wife and I and Betty Michell, her husband being gone to Westminster. Here at church (God forgive me), my mind did courir upon Betty Michell, so that I do *hazer con mi cosa in la eglisa meme* [so that I do play with my thing even in the church]'.[2]

Pepys did not seem to have very much guilt about this activity, or apparently much fear of being discovered. Perhaps this was the golden age of masturbation – a time of relative innocence after the moral strictures of the church had begun to fade, but before the Enlightenment replaced them with a new, and just as condemnatory, authority of 'science'. Because just a few years after Pepys's death in

1703, a tract was published that heralded a crushing attitude to masturbation that would last 250 years.

The publication, around 1712, was called *Onania; or the Heinous Sin of Self-Pollution*.[3] Its full title is shown in Figure 22: Onan is dragged into it, even though he apparently only practised withdrawal or 'coitus interruptus'.[†]

This anonymous work wanted to 'expose a sin so displeasing to God', but mainly revelled in descriptions of the physical effects of self-pollution, particularly for young men, 'who were strong and lusty before they gave themselves over to this vice, have been worn out by it, and by its robbing the Body of its balmy and vital Moisture, without Cough or Spitting, dry and emaciated, sent to their Graves'. Girls were at particular risk, as the author's friend had been assured by a governess of an eminent boarding-school 'with tears in her eyes, that she had surpris'd and detected some of her Scholars in the very Fact; and who upon Examination confess'd, that they very frequently practis'd it, cum Digitis & aliis Instrumentis, and that chiefly those of them from the Age of about 15 and upwards'.

Thomas Laqueur, author of the definitive history of masturbation,[‡] has claimed that *Onania* was written by John Marten, a surgeon with the dubious prior record both of having introduced the term 'masturbation' into English, and of being prosecuted for obscenity. He was an imaginative businessman – the pamphlet finished with advertisements

---

† 'And Judah said unto Onan, Go in unto thy brother's wife, and marry her, and raise up seed to thy brother. And Onan knew that the seed should not be his; and it came to pass, when he went in unto his brother's wife, that he spilled it on the ground, lest that he should give seed to his brother. And the thing which he did displeased the LORD: wherefore he slew him also.' Genesis, Chapter 38.
‡ *Solitary Sex: A Cultural Theory of Masturbation* – this work, more than 500 small-print pages of erudite scholarship and footnotes, has unfortunately led the venerable Laqueur to become known as 'Professor Wank'.[4] Let's hope this is not the fate of all academics that venture into this area.

for 'Strengthening Tincture' at half a guinea a bottle and a 'Prolific Powder' at 12 shillings for 24 doses, very expensive items which could be bought from the bookseller Mr Crouch in Paternoster Row.† *Onania* started an industry in such texts, with the eminent Swiss physician Samuel-Auguste Tissot's *Onanism, or, a Treatise upon the Disorders produced by Masturbation* becoming one of the best-sellers of the 1700s and one of the most influential books in medical history.[5]

Tissot was obsessed with the dangers of loss of semen, citing past authorities that it brought on 'on a lassitude, a feebleness, a weakening of motion, fits, wasting, dryness, fevers, aching of the cerebral membranes, obscuring of the senses, and above all the eyes, a decay of the spinal cord, a fatuity, and other like evils'. So the idea that masturbation makes you go blind has a long history. But Tissot does provide one statistic: a suspiciously accurate, and much-quoted, assessment that 'losing one ounce of sperm is more debilitating than losing forty ounces of blood'.‡ So the moralistic religious censure of non-procreative sex morphed easily into a pseudo-scientific justification.

Some statistics finally started to appear in William Acton's *The Functions and Disorders of the Reproductive Organs* in 1857, which says that out of 1,000 male patients with 'phthisis' (tuberculosis), 12% had committed sexual excesses, 18% had

----

†Next to St Paul's Cathedral, which later became the centre of the bookselling and publishing trade until it was utterly destroyed, along with 5 million books, by incendiary bombs on 29 December 1940.
‡I can't resist looking at this. An ejaculation is about 3mls, which according to Tissot is equivalent to losing forty times as much blood, or 120 mls. The adult male body contains around 5 litres of blood, so 120 mls is around a fortieth of a body-full. So according to Tissot a man has got forty orgasms' worth of energy inside him. If you donate blood, they will take around 500 mls, that's four Tissot-orgasms, but after giving blood it apparently takes several weeks for the red cells to be replaced, whereas the recovery from four orgasms should be somewhat quicker.

been addicted to masturbation, and 22% had suffered from 'involuntary emissions'.[6] We should, however, give Acton credit for making allowance for what is now known as 'reverse causation': 'delicate constitutions [...] may be more susceptible to sexual excitement'. This insight was missed by all the doctors who observed that patients in mental asylums masturbated, and assumed that it was this habit that brought them there in the first place.[†]

In America, the Seventh-Day Adventist and medical doctor John Harvey Kellogg, celebrated inventor of the corn flake, expounded on the vast range of medical effects of self-abuse, from piles to indigestion and constipation and, of course, dimness of vision and insanity. Urges could be controlled by fifteen minutes of bathing the genitals in cold water before retiring, and eating unstimulating food. Such as breakfast cereal. Graham Crackers were similarly born, or at least commercially promoted, through their anti-masturbation properties. And there was also a wide range of patented mechanical and electrical devices to stop the practice, too awful to contemplate: designed to prevent masturbation or nocturnal erections through electric shocks, or spikes on the inside of a tube to be worn over the penis.[7]

However, just as the medical world started to realise in the late 1800s that 'self-pollution' did not inevitably lead to the asylum, the 'social purity' movement really got going. Its adherents were determined to increase sexual continence, and particular targets for 'purity' were English public-school boys, who were seen as the future leaders of the British empire.[8] There was an old fear of nurses 'soothing'

---

†Acton warns us all to be on the look-out for 'the characteristic sunken eye of the masturbator', and cites Jean-Jacques Rousseau's *Confessions* as a 'tolerably accurate portrait of a masturbator half-way on the road to ruin', but displaying 'the unmanliness, the pettish feminine temper and conceit, which would make a hearty English lad shudder with disgust'. So much for the French.

their charges to help them to sleep, masturbation being (correctly) recognised as a fine soporific. Another aspect of this pre-Freudian view of children being asexual beings was the popular image of corruption by an external influence: historian Alan Hunt reports the ex-head of Metropolitan Police CID telling a purity rally in 1910 of

> a harrowing story of an Eton boy, son of a colonel in the army, a brilliant lad, always head of his class [...] who had been reduced to drivelling imbecility as the result of secret sin, induced by the sight of an obscene photograph exhibited by a scoundrel whom he met in a railway train. I had the satisfaction of hunting the villain down and of procuring him a long sentence of penal servitude.[9]

That must have been quite a dirty picture.

America did not have much of an empire to worry about, but there was still concern about its youth. So in 1898 the Young Men's Christian Association (YMCA) sent out questionnaires and got 251 replies from students in 75 colleges and theological seminaries – sadly we don't know the response rate. The respondents' average age was 23½, and they were nearly all Christians. When asked 'What was your severest temptation of school days?', 132 admitted 'Masturbation', to which 131 'yielded', and 69 admitted to yielding 'much'.

This is hardly a random sample, but 131 out of the 232 who answered this question is 57%. The author could not 'fail to be impressed with the alarming extent to which some students practise masturbation', and in particular that 75 were guilty of this vice after their conversion to Christianity, and 24 even after they decided to become ministers. This survey was published in 1902 by G. Stanley Hall, and has been credited as the first published sex survey.[10] Hall was very eminent, a friend of Freud and one of the founders of American psychology, who wrote of 'Masturbator's Heart' and other diseases brought on by 'solitary excess', and who, when a boy, had 'taped down his penis to avoid unconsciously produced erections.'[11]

The irony is that the medical profession had been giving 'pelvic massage' to women suffering from 'hysteria' for years, in order to bring them to a 'hysterical paroxysm'. A range of vibrators had been developed for use in the surgery and the home to help in this tiring task of producing orgasms: the steam-powered model was fortunately superseded by an electric version in 1902, ten years before the vacuum cleaner was electrified. This encouraged home use: a 1914 US advert for the 'Premier Vibrator' ('Gives both Pounding and Rotary Strokes') suggested to men that they should 'buy one for your wife or sister. Maybe she'll get a chance to use it – when you're not at home.'[12]

The panic starts to fade

2.7%: the proportion in Kinsey's male sample who claimed to have successfully self-fellated[13]

Anti-masturbation rhetoric reached its climax in the years preceding the First World War, but new authorities had begun to speak up: not yet in favour of the practice, but eventually concluding it was neither morally wrong nor physically dangerous. Few ordinary people read the weighty tomes of Havelock Ellis, whom we met in the last chapter discussing 'inverts', but he started to influence professionals by claiming in 1899 that 'auto-eroticism' was normal and not a medical problem.[14] Soon after, Magnus Hirschfeld, always happy to court controversy, brought more criticism on his head by proclaiming that 96% of men had masturbated.[15]

In their interviews with people born between 1901 and 1931, Szreter and Fisher found men and women had rather different recollections.[16] All the men knew about masturbation, none denied they had done it, and they viewed it as commonplace and a 'natural' part of growing up. But out of 57 women, 7 said they had not known there was such a thing (and a few still did not know), and only 2 were prepared to admit they had done it. 'Antonia' (born in 1928) said 'you didn't talk about things like that. You didn't do things like that. It was all a bit nasty really.' Szreter and Fisher conclude that the females prized their innocence and so did not make very much effort to find out more about sexuality.

The contrast with Katherine Davis's contemporary 1929 survey of 2,200 US women could not be greater. We've already seen that this highly selected, highly educated group reported high rates of same-sex behaviour, and Davis was also keen to promote liberal views on female masturbation; she recorded that 60% of unmarried respondents had masturbated, and 30% of married respondents had done so before marriage, although two out of three of these women considered the habit 'morally degrading'.[17]

It should not come as a surprise that Kinsey had strong views about masturbation. He had been a Boy Scout before the First World War, when the official line was 'Any habit which a boy has that causes this [sex] fluid to be discharged from the body tends to weaken his strength.'[18] But cold showers every morning could stop neither his habit nor his consuming guilt. He later crusaded against this dogma of blame and shame, pointing out that as late as 1940 the United States Naval Academy at Annapolis still ruled that a candidate 'shall be rejected by the examining surgeon for evidence of masturbation'. He was also deeply sceptical, as he was of all 'theoretical' psychoanalysis, of the European, Freudian view that female masturbation was just an infantile substitute for proper intercourse.

One of Kinsey's headline statistics was that 92 to 96% of men have experience of masturbation, with the higher figure for the more educated (incidentally, right on Hirschfeld's ridiculed estimate). So Kinsey claimed that 'self-abuse', just like other 'unnatural' elements in the sexual repertoire, such as oral sex, was not just a characteristic of the lower classes. Around 87% of unmarried 16- to 20-year-old men reported masturbating – slightly less in devout religious groups – representing 48% of their 'total outlet'. And marriage did not necessarily mean the end of the practice: 22% of married 36- to 40-year-olds masturbated, making up 13% of their total outlet.[19]

Just like the questions about anal sex, some of his more imaginative statistics were not made public until 1979, when the summary tables were finally published.[20] Subjects were asked, 'How much of the masturbation is by inserting something into the penis?', a question that may have arisen out of Kinsey's own interest in this masochistic practice.[†] Out of over 3,000 white college men (the only sample from Kinsey that might be moderately reliable), 9% said they had tried it and 1% used this technique much of the time.

The next question, presumably asked of the three statisticians as of everyone else, was 'Were you able to put your mouth on your penis?'. If the answer was 'No', they were asked 'Did you try?'; if 'No', the interrogation did not stop but persisted with 'Did you think about trying?' Responses to this unusual query about self-fellatio revealed 15% who had thought about it but never tried, 26% had tried and failed, and a flexible 2.7% (around 100 out of 3,700) claimed they had succeeded. A round of applause, please.

When it came to females, Kinsey reported that 60% of women had masturbated at some point, the same as Davis had found twenty years earlier in her equally non-random sample.[22] Always interested in extremes, Kinsey found four

---

[†] Kinsey experimented with masturbating with objects inside his urethra, including pencils and eventually a toothbrush, brush end downwards.[21]

women who reported doing it at least 30 times in a week, but the median frequency in young single women was once every 2½ weeks, and once every five weeks in young married women. As we'll see below, this does not seem far from current British practice.

44%: the proportion of British women aged between 25 and 34 who had masturbated in the previous four weeks (Natsal-3, 3*)

The increasing normalisation of masturbation continued with Masters and Johnson,[†] who provided vivid descriptions of 'auto-manipulation' techniques for both males and females.[23] But the 'phallic fallacy' of the dangers of excessive masturbation was still powerful in the early 1960s. Their sample of 312 men had a wide range of habits, but in all cases the males considered as 'excessive' whatever was greater than their current habit: the once-a-month man was concerned that once or twice a week would lead to mental illness, while the individual who performed two or three times a day thought that five or six times a day might lead to 'a case of nerves'.

In the write-in surveys of the 1970s Hite reported that 82% of her respondents currently masturbated,[24] and 72% of the *Cosmopolitan* survey respondents said they did it at least several times a month.[25] The 1992 US National Health and Social Life Survey (NHSLS)[26] prefaced their questions on masturbation with the statement 'Masturbation is a very common practice', which might be thought to have influenced the response, but got very different answers to both

---

† William Masters and Virginia Johnson are famous for their laboratory work on sexual response, starting in the late 1950s, in which they studied first 145 prostitutes and then 312 men and 382 women recruited from the local community. While making a large contribution to the understanding of physiology and sexual response, any statistical conclusions are limited, owing to their highly selected sample.

Figure 23: **Proportion reporting masturbation in the previous four weeks**

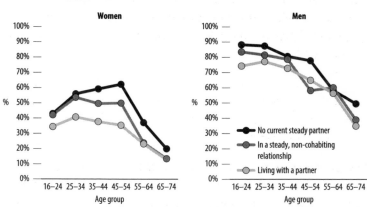

Source: Natsal-3 in 2010.

Hite and *Cosmo*, and even Kinsey: only 60% of men and 40% of women reported having masturbated in the last year. Even then, they said that the 'large number of adults who masturbate seems puzzling', and were clearly surprised that there was little relationship with age, and rates were apparently higher in people with a partner.

But what about Britain, birthplace of onanism? The first Natsal survey, having struggled to get funded, had intended to include questions on masturbation but gave up when developing the questionnaire 'because discussion of this practice had met with both distaste and embarrassment from respondents'.[27] But they persevered in Natsal-2[28] and Natsal-3,[29] asking 'When, if ever, was the last occasion you masturbated? That is aroused yourself sexually', and providing a range of responses, from 'in the last seven days' to 'never'.

The proportions in the previous four weeks reported in Natsal-3 in 2010 are shown in Figure 23, shown separately for those living with a partner, those in a steady but non-cohabiting relationship and those with no steady partner. When we combine these groups, 66% of men and 33% of women reported having masturbated in the previous four weeks. There is a declining rate with age, particularly for men, when rates went down from 83% in 16- to 24-year-olds to 33% in 65- to 74-year-olds: the maximum rate for women was 44% in the 25- to 34-year-old group.

The proportion who reported recent masturbation is higher in those with no current partner, but not by a major amount, particularly for men. But the gender difference is clear – there is roughly twice the rate among men as among women, although there may be some under-reporting in women because of stigma still attached to the practice.

The same question was asked ten years previously, in the Natsal-2 survey, which only asked people in the age range 16–44. There was an almost identical pattern by age, but with a slightly lower rate: this may be a genuine change in behaviour, but it is more likely that the intervening decade has made it more socially acceptable to acknowledge masturbation. Closer analysis of the Natsal-2 data reveals higher rates in those with higher education, and lower rates among the more religious – exactly the findings of Kinsey.

Perhaps the most interesting finding in Natsal-2 was that, for women, more sexual activity was associated with increased masturbation, while the opposite association held for men: the authors concluded that 'it is difficult to avoid the conclusion that masturbation for many predominantly heterosexual men may represent a substitute for vaginal sex, while for women the practice appears to be part of the wider repertoire of sexual fulfilment, supplementing, rather than compensating for, partnered sex among women.'[30]

Attitudes have certainly changed. In response to current rates of teenage pregnancy and sexually transmitted diseases in young people, masturbation has come to be seen by some as a positive aspect of sexual health, encouraging awareness of sexuality in solitary safety and – now that blindness, insanity, clammy hands and pimples have finally been ruled out as side-effects – providing pleasure without associated risks. A recent (2*) survey found 46% of US women between 18 and 60 had used a vibrator for masturbation, 20% in the previous month.[†31] And a study of online sex-aid retailer Lovehoney.co.uk revealed that vibrators comprised 18% of the products sold, around 400,000 a year.[32]

But a surprising amount of social stigma remains: when Joycelyn Elders, appointed by President Bill Clinton as the first black US Surgeon General, was asked at an AIDS conference in 1994 whether masturbation could be discussed as a safe-sex activity, she replied: 'I think that it is part of human sexuality, and perhaps it should be taught.' She was sacked ten days later by Clinton. And not everyone was delighted when, in 2009, the NHS in Sheffield produced a leaflet entitled *Pleasure* that dared to suggest that teenagers had sex because it was enjoyable. It carried the slogan 'an orgasm a day keeps the doctor away', recommended regular masturbation for teenagers and got a poor reception from the *Daily Mail*.[33] Remnants of the taboo are still apparent three centuries after *Onania* hit the shelves.

Not all is rosy. The rise in internet pornography has led to renewed concern about compulsive masturbation. In a parallel to despairing letters of a century ago, men who feel they are addicted to pornography compare experiences on websites such as yourbrainonporn.com or nofap.org:[‡34] a particular concern is erectile dysfunction from excessive masturbation. 'Fapstonauts' set themselves challenges to last, say 90 days, without a 'reset' by succumbing to porn/masturbation/orgasm (PMO): posters to the Forum record they have managed a streak of '40 days since I last PMO'd',

---

† This was based on a market research panel with a 54% response rate. The study was funded by the makers of Trojan condoms.
‡ 'Fap' being internet slang for males, 'schlick' for females, both onomatopoeic terms based on the sounds made by masturbation.

and get supported in their efforts to 're-boot' themselves to manage without artificial stimulation.

And some really seem to be addicted. A recent study in Cambridge recruited 19 men aged between 19 and 24 with confirmed 'compulsive sexual behaviour', found through internet-based advertisements and referrals from therapists. Both they and a control group were exposed to pornography while in a brain scanner: not only were those with 'compulsive sexual behaviour' more responsive, but the parts of the brain that were activated were part of the 'reward network' that is also activated in drug addicts.[35] Martin Daubney, the former editor of lads' mag *Loaded*, has said that easy access to internet porn is 'like leaving heroin around the house'.[36]

But although a small minority may suffer a 'PMO' compulsion, there seems a risk of turning a reasonably common behaviour, such as watching pornography, into a medical disorder. Although it's easy to laugh at the historical excesses of the 'great masturbation panic', concern about the practice has turned out to be a remarkably resilient phenomenon.

7

HOW IT ALL STARTS

If you were a teenager 500 years ago, your dating prospects were bleak. Whatever their class, in their early teens many were sent away to work for another family for up to ten years with scant opportunity – or time – to mingle with potential partners. Of course, we've got to assume that young people still had some sexual interest in each other, but this was closely policed by society, whose obsession with young people's sexual activity was clearly under way in the early 1700s, when the perils of masturbation in schools started to be proclaimed. And perhaps they had reason to be worried: as the eighteenth century progressed, illegitimacy rates rose, people got married younger and a bigger proportion of brides were pregnant when they married. The young were becoming more sexual, and this has always been seen as a threat.

Jump ahead a couple of hundred years, and we find the rise of youth culture in the 1950s. This saw an explicit and precocious teenage sexuality that conflicted with the spartan wartime ethic of their parents – Elvis really did move his pelvis in a provocative way. With the invention of the teenager came the 'teenage problem', and the 'generation gap' (if you are old enough to remember this popular media term). The rise of teenage pregnancy in the 1960s and 1970s, accompanied by sexually transmitted diseases, brought funding for studies into what kids were actually doing, with a particular concern about when they finally went 'all the way', the full Monty, got to 'number 10' and other euphemisms for full sex.

It's difficult enough to ask adults about their sex lives, but trying to find out what adolescents get up to is a serious challenge for researchers. Should you ask adults what they did when they were younger? They may be willing to answer, but how far we can trust their memories is another matter. So perhaps we should ask the young people themselves – but will they tell the truth? Official statistics might give some clue, but we can only study the consequences – teenage pregnancy – or check on the use of contraception services.

Let's start with the obvious question: how old were you when you first had sex?

Age at first sex

16: the median age at which people first
have sex in Britain (Natsal-3, 3*)

If it's happened, you can probably remember the first time – such life events tend to stick in the mind. But how do you stack up against everyone else?

Figure 24 shows the proportion of different 'cohorts' of people who had had sex by each age.[†] To use this plot, find the curve that corresponds to the period when you were born. Then remember the age you first had sex, and run your finger upwards from that age until you come to 'your' line.

---

† There is some statistical sophistication in producing these curves, as teens who have not yet had sex only contribute information up to their current age: these 'survival analysis' techniques are also used when estimating likely lengths of marriages.

118

Figure 24: **For people born in a particular period, the percentage who report having had sex by each age**

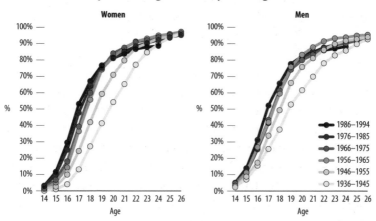

Source: Natsal-3 2010.

Then run across to find at what 'percentile' you lie. If it is, say 80%, this means that 80% of your cohort reported having sex earlier than you, and you started comparatively late. If you cross at the 5% point, you were in the vanguard of your generation.

Note that for each successive cohort the line shifts to the left, as first sex tends to occur at a younger age. By looking at the age the curves cross 50%, we can also read off the median age at which different groups reported first having sex: for men and women born around 1930, the median age was 18 for men and 19 for women, and so occurred in the late 1940s. For those born around 1980, the median age at first sex was 16 for both men and women – only a few years' reduction, but a very important few years.

There's a particular interest in the proportion of adolescents who have sex below 16 – this is the official British 'age of consent' for both females and males, and so, strictly speaking, this is illegal even if both participants are below 16.

Figure 25 shows the percentage reporting that they had had sex before 16, for cohorts of people born in ten-year

Figure 25: **The percentage of each cohort who reported having sex before age 16**

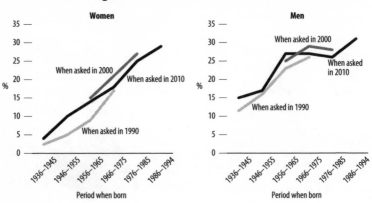

The three lines correspond to the three surveys, Natsal-1 in 1990, Natsal-2 in 2000 and Natsal-3 in 2010. If people answered accurately, the lines should roughly lie on top of each other.

age bands. Looking at the line for the 2010 survey shows that there has been a huge increase with successive generations. Around 1 in 7 men born around 1950 (my generation) reported having sex before 16, and around 1 in 10 women – the event would have occurred in the late 1960s, when sexual behaviour was starting to warm up. This proportion then steadily grew through the decades until, for those born around 1990, the figure is close to 1 in 3 for both men and women – the gender gap has closed.

Members of each cohort have been questioned at a number of surveys as they have got older: for example, those born between 1956 and 1965 were questioned in 1990, when they were around 30, in 2000, when they were around 40, and in 2010, when they were around 50. As the age at which you first have sex does not change as you get older, strictly speaking we would expect the lines to be roughly the same: just as when we asked about same-sex experience in Chapter 5, even though they are not actually the same individuals, we would expect the proportions only to vary owing to the play of chance influencing who happened to be picked.

Fortunately the lines are fairly similar, although the reports from women, and to a lesser extent men, show some increase between 1990 and 2000, and the Natsal team put this down to more honest reporting in 2000 as sexual attitudes became more relaxed and better interviewing techniques were used.[1]

The 'one in three' figure doesn't mean that in every Year 11 class, when school students become 16, a third have already had sex. The rates are lower in more affluent areas, but higher in more deprived – we shall see the consequences of this when we come to teenage pregnancy.

5%: the proportion of girls born around 1945 who had had sex before they were 16 (Natsal-3, 3*)

The data we've just seen suggest that asking people what happened when they were teenagers does not always end up with the correct answers – perhaps some events would rather be forgotten. So what about asking the kids themselves what they are getting up to?

The first serious attempt to do this in the UK was carried out by Michael Schofield in the 1960s and resulted in the popular paperback *The Sexual Behaviour of Young People*.†[3] His team drew up lists of 15- to 19-year-olds from selected areas, and then tried to interview a random sample of around 2,200. After sending introductory letters, the interviewers made up to six house calls before giving up, often dealing with parents who could be rude or aggressive. After 'repeated calls on cold wet nights at houses where they could not expect

---

† This was funded by the Nuffield Foundation. Schofield was a brave choice to lead the project – he was an ex-Second World War RAF fighter pilot but was gay, and had written books that took a positive view of homosexuality under the pseudonym (due to homosexuality still being illegal) of Gordon Westwood. He died, aged 94, in 2014, after a lifetime of effective campaigning for civil liberties.[2]

to be welcome', they achieved an outstanding 85% response rate. Sex researchers always face the risk of press attacks, but these interviewers were particularly courageous; the team was relieved when they managed to finish the survey without a public outcry.

It was only a decade after Kinsey, and so they followed his interviewing style: questions were asked rapidly, in a flexible order, using vernacular terms. Each child was asked what 'stage' they had reached, starting at dating, kissing, deep kissing, touching breast over clothes, then under clothes and so on and so on, up to full intercourse (unlike more recent surveys, they did not include oral sex as one of the stages – it is simply reported as 'very rare'). Schofield found, for example, that at age 15 less than a fifth of girls had had their breasts touched under clothes, but by 17 over half had experienced this. The headline statistic was that only 14% of boys and 5% girls had had sex by age 16 – hardly the hotbed of adolescent depravity that was feared. These figures, for a generation born around 1945, fit reassuringly with the Natsal results given in Figure 24.

78%: The proportion of US adolescents who, when re-interviewed, changed their mind about the month they first had sex (2*)

We've already noted that a quick and easy way to do surveys is by using online panels that have already been recruited by market research companies. For example, in 2008 YouGov conducted a poll of over 1,400 14- to 17-year-olds for Channel 4 in which 22% said they had had sex and 15% did not answer: Channel 4 then reported this as the great 'Teen Sex Survey', and claimed '40% of all 14- to 17-year-olds are sexually active', a classic example of media mis-reporting of sex research.[4] We shall see a lot more. The panel comes over as being, if anything, rather moderate: the original report from YouGov shows that far more were concerned about their weight (46%) than about anything to do with sex, and of those that were sexually active, only 8% said they had had sex after drinking too much.[5] But Channel 4 managed, somehow, to turn this into '35% of sexually active teens have had sex after drinking too much'. Complete o* nonsense, and damaging too.

But can we ever believe a word of what kids say? Fortunately, repeat surveys of the same individuals allow a check of whether they change their minds about what they have done. In the US National Longitudinal Study of Adolescent Health, 5,000 young people were interviewed at an average of 15½ and were asked if they had had sex and, if so, what age they had first done so, and then the same question was put 18 months later to exactly the same people.[6] They should have given the same answers, but didn't.

Overall, of those who said they were sexually active at the first interview, 11% denied this when asked again, essentially reclaiming their virginity at the first interview. On average, respondents moved their age at first sex to 4½ months older, although white girls were the most consistent, with a mean shift of only 2½ months. The researchers do not rule out the idea that much of this may be a simple inability to remember dates, and a few months were not so important anyway, but it's a warning that maybe adolescents are not the most reliable of witnesses. And their 'pledges' of virginity may not be reliable either.

82%: the proportion of US adolescents pledging virginity before marriage who later denied they made the pledge (2*)

Surveys of adolescents are now routine around the world, and are generally carried out in schools rather than at home. They make the Schofield study seem like an age of comparative innocence.

For example, the annual US Youth Risk Behavior Surveillance System asks high-school students aged 14–18 about all the things that they would normally want to keep under wraps, such as taking drugs, carrying guns and eating vegetables.[17] In 2013, 47% of students had had sex, ranging from 30% in Grade 9, aged around 14–15, to 64% in Grade 12, when they were 17–18, a rate that has stayed fairly stable since 2001. Sexual activity used to be even higher: it was 54% in 1991, and the median age at first sex reached a low of 17 in 1978.

So does this decline in sexual intercourse mean that American teens are sitting round playing video games, or are so distracted by social networks that they have lost interest in each other's bodies? It seems rather implausible, and raises the question of whether oral sex is being used as a substitute activity. Indeed the National Survey of Family Growth found that around half of sexually active young people had oral sex on its own before having vaginal intercourse – the other half jumped straight to the business.[9] And in a US study of 580 14-year-olds, oral sex was considered safer and more acceptable.[10]

Perhaps the 'technical virginity' of sticking to oral activity is an inevitable, and possibly quite reasonable, consequence of encouraging teenagers to delay having full sex. At one end of the spectrum this can simply take the form of encouraging teenagers to wait until they feel emotionally ready, but a controversial alternative is the American phenomenon of 'pledges' to abstain from sex before marriage. Christian organisations such as 'True Love Waits' promote their pledge‡ amid a variety of other online merchandise

---

† The 2013 survey found that 9% of males in Grades 9–12 (aged 14–18) had carried a gun during the previous month (21% in Arkansas), and 7% had not eaten vegetables in the previous week (excluding French fries).[8] It was based on over 13,600 responses: 77% of schools agreed to participate, and 88% of students in those schools.

‡ True Love Waits pledge (2009): 'I am making a commitment to myself, my family, and my Creator, that I will abstain from sexual

such as a Purity Solitaire Heart ring for only $25.95.[†11] But do these pledges work? This is a deeply divisive issue in the US: they were heavily supported by President George W. Bush's administration, and by 2008 US$200 million a year were being spent on such programmes.

Just as with anything that people choose to do, the 'effect' of pledging is very difficult to research: we cannot randomly assign people to pledge or not, and then follow them up to see who has sex first. A study that compared pledgers and non-pledgers found that the 289 pledgers did appear to delay having sex by 2 years, but 88% of the pledgers did have vaginal sex before they got married. And many of the remainder had had oral sex, which (unless you follow Bill Clinton's example) hardly seems to fulfil the pledge to abstain from sexual activity.[12]

And maybe the sort of people who would sign a pledge would have delayed starting sex anyway? Pledgers tend, unsurprisingly, to be more religious, are more negative about sex and have less sexual experience than non-pledgers. A more sophisticated analysis used the same dataset but matched each of the 289 pledgers with around 3 similar students with similar characteristics on religiosity and other factors that are associated with pledging.[13] Five years later the two groups had indistinguishable rates of vaginal, oral and anal sex – the only difference was that pledgers used less birth control.[‡]

---

activity of any kind before marriage. I will keep my body and my thoughts pure as I trust in God's perfect plan for my life.'

†Miley Cyrus used to wear such a ring, but it has not been in evidence in her recent videos.

‡This re-analysis used the technique of 'propensity scores', which creates groups of individuals matched in the factors that are associated with pledging, and then within each group compares the outcomes of those that did pledge with those that did not. In Chapter 13 we will see that a similar type of analysis has been used to take apart claims about the 'effect' on future behaviour of watching sexual content on TV.

But perhaps the most notable finding is that, five years after their solemn pledge, 82% of pledgers denied ever having made it. This should not be surprising, as by then around this proportion were having sex, and they clearly could not cope with a high degree of 'cognitive dissonance'. The researchers conclude that this non-binding nature of the pledge means that sex education has to be provided to everyone – perhaps especially pledgers.

But the argument continues: Barack Obama was keen to drop funding for pledging programmes but was forced to put some back in to get his Obama-care bill through.[14]

***

71%: the proportion of 15-year-old girls in Greenland who report being sexually active

***

Britain has followed the general pattern in richer countries: there has been a general decrease in median age at first sex, and then it has stabilised at around 16. But there is a wide variety of practice around the world. The Health Behaviour in School-Aged Children (HBSC) survey estimates the proportion of 15-year-olds who have had sex in each of 43 countries, nearly all of them in Europe.[†15] The highest rates are in Greenland, where 71% of girls and 46% of boys report having sex when aged 15, with the next highest being their 'mother country', Denmark, where the proportion is 38% of both girls and boys. For England they report 32% of girls and 26% of boys, just about matching the Natsal results.

Countries with more traditional cultures can show

_____

† The Health Behaviour in School-Aged Children survey collects data every four years on around 1,500 young people aged 11, 13, and 15 in each of 43 countries: 200,000 altogether. Questionnaires are completed by students in the classroom, with over 60% response rates in most countries. It is a World Health Organisation collaboration. I would rate it as 2* or 3*, depending on the country.

substantial gender disparity: in Greece 18% of girls but 39% of boys report having sex when aged 15. Perhaps the most interesting statistic is for the Netherlands, with its open approach to the discussion of adolescent sexuality, where relatively low proportions of 22% of girls and 19% of boys report sex when aged 15: around two-thirds the rates in the UK. As we shall see later, there is an even bigger disparity between the Netherlands and the UK in teenage births.

56%: the proportion of women in Natsal-2 who were 'not competent' when they first had sex

---

So what was it like for you – you know, the first time? Was it crisp white sheets with the gentle and caring soulmate of your life? A moonlit beach on a Greek island in the throes of a holiday romance? Or was it a quick business on top of a pile of clothes in a back room at a party?

And how competent were you? And I don't mean did you know how to deal with the appropriately arousing bits of body (which is very doubtful), or whether you could open the condom packet without dropping the contents into the sand. I mean officially 'competent', in the sense that the Natsal team use. Would you agree with any of the following four statements?

1. One of us was more willing than the other.
2. I wish I had waited longer.
3. The main reason was peer pressure or because I was drunk or had taken drugs.
4. I did not use reliable contraception.

A single 'yes' means that your first sex is considered 'not competent' by Natsal. So now you have a label, and you can compare yourself with the rest of the population.

Of 16- to 24-year-olds who were born around 1980, having

first sex in the late 1990s, 56% of women and 46% of men were 'not competent' at first sex.[16] This rose to 91% of women and 67% of men if first sex happened when aged 13 or 14, but 37% of women and 39% of men were 'not competent' even when their first sex was when aged 18 or over. We can, of course, look at these numbers in a rather more positive way: 44% of women and 54% of men *were* competent at first sex.

There are gender differences. Natsal report that 42% of women, but only 20% of men, regretted the timing. And 22% of women said their partner was more willing, and only 7% of men. But there was much closer agreement about whether contraception was not used (around 10%), and about whether it happened because of being drunk or owing to peer pressure (around 16%). There is some social anxiety about young girls being pressured into sex before they are really ready, but these statistics reveal that there is also a reasonable minority of males who have mixed feelings about what took place.

Competence is increasing across generations: of women born around 1948, only 32% said they fulfilled the criteria for 'competence' when they experienced their first sex in the 1960s. But for the generation born around 1982, 55% were competent when having their first sex in the late 1990s. The current generations seem even more savvy, and competence at first sex is not just a matter of embarrassing memories: Natsal-3 data suggest that it is an independent predictor of subsequent sexual health problems.[17] So it's not just *when* sex starts, but *how* it starts, that relates to future difficulties: bad experience of sex can shape our future world. And this suggests that sex education should focus not just on the mechanics of sex and contraception, but also on the 'stages' of sexual activity and the appropriateness of taking each step.

In America, too, young people are increasingly competent: when 18- to 24-year-olds were interviewed for the National Survey of Family Growth in 2006–10, 11% of women and 5% of men described their first sex as 'unwanted', compared with 13% of women and 10% of men a few years earlier, in 2002.[18] And 70% of females and 56% of males said it was with a steady partner. But note the mismatch between the sexes in these proportions. Have some of these 70% of girls been cruelly deceived – did they think they were with a steady partner, when the males thought they were with just a friend? Or are females losing their virginity to sexually experienced men who are steady partners, whereas more men lose it to a more casual acquaintance?

But for either gender, the really casual acquaintance is rare: in Natsal-2 only 1 in 20 reported having first sex with someone that they had just met.

62%: the proportion of sexually active 15-year-old girls in Germany who report being on the pill

A crucial aspect of 'competence' is the use of effective contraception, but this means that contraception has to be available to young people. It may be difficult for a current generation to grasp how controversial this used to be. When, in 1965, Michael Schofield made the tentative suggestion of making birth control available to adolescents, the *British Medical Journal* responded by saying that 'To give official sanction to such facilities for the unmarried must inevitably promote extramarital and promiscuous sexual activity.'[19] But by this time the Sixties were in full swing, the 'unmarried' were getting on with their 'activity' regardless of the availability of contraception services or the pompous utterances of the *British Medical Journal*, and there was a predictable surge in teenage pregnancies and sexually transmitted diseases.

But things have improved. Of those who had their first sex in the 1960s, only 65% claimed they used contraception the first time, compared with 93% who had first sex in the late 1990s. In the USA, in 1982 less than half of teenagers reported using contraception at first sex, compared with around 80% in 2006–10.[20] And the Health Behaviour in School-Aged Children survey reported high usage of contraception in 15-year-olds having sex: more than 85% reported using a condom at their last sex in countries as diverse as Estonia, Greece, France and Slovenia, while in Germany, Denmark and the Netherlands more than half the sexually active 15-year-old girls were on the pill, reflecting their developed sexual health services for young people.[21]

5%: in 1972, the proportion of 16-year-old girls in England and Wales who got pregnant

2½%: in 2012, the proportion of 16-year-old girls in England and Wales who got pregnant

---

Except in rare circumstances, when a girl in her mid-teens gets pregnant, something has gone wrong. But it happens: in 2012 one in forty 16-year-olds got pregnant in England and Wales, of which half went on to have their baby.

The numbers of teenagers who give birth each year is a well-used statistical marker of changes in society, and Figure 26 shows the birth rate of 15- to 19-year-olds since 1938. There are generations of social history in this single graph; before the war, lack of access to contraception, combined with a social stigma attached to illegitimate births, resulted in a restraint in sexual behaviour that meant that less than 1 in 50 of all 15- to 19-year-olds gave birth each year. Then, as the atmosphere became more liberal, there was a steady increase throughout the 1950s and 1960s, to a peak of 1 in 20 in 1971, still mainly within marriage (though those marriages were generally precipitated by the pregnancy). And as contraception and abortion became more available, there was a precipitous decline to a steady rate of around 1 in 30 a year, with a recent decline to levels not seen since 1947. And

Figure 26: **Percentage of women aged 15–19 giving birth each year 1938–2012, England and Wales**

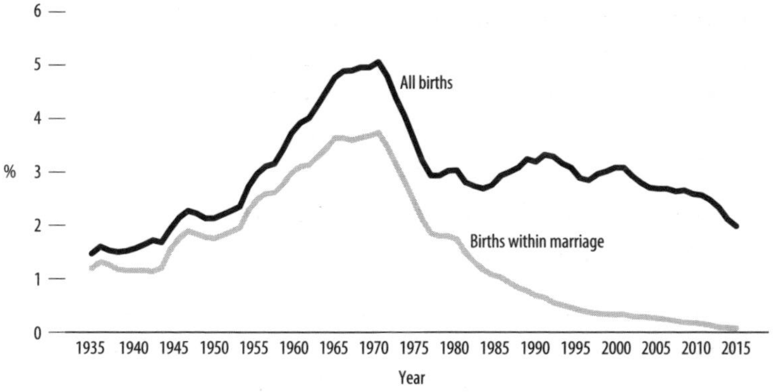

Source: Office for National Statistics.

the main change over the last 30 years is that now almost none of these births is to a married mother.

But not every pregnancy results in a baby, and if we want to learn about sexual *behaviour* – then we really need to look at conceptions as well as births. Fortunately, the Office for National Statistics estimate the number of conceptions by adding births to known abortions (though this means that conceptions ending in miscarriages or illegal abortions are therefore missed).[22]

The official teenage pregnancy rate is defined as the proportion of girls aged 15–17 that get pregnant each year. In 1998 this was 4.7%, around 1 in 20, and a bold target was set to halve this rate by 2010. Figure 27 shows that not much happened until 2007, and then a steady drop reduced the rate to 2.8% by 2012, the lowest since 1969. This is a 41% relative reduction since 1998 – not quite meeting the target of halving the rate, but still a great achievement.†

So why has this happened? Have teenagers stopped

---

† In 2012, 1,282 14-year-olds got pregnant, and 253 13-year-olds. Half of all conceptions in under-18s led to abortion, rising to two-thirds of 14-year-old pregnancies and three-quarters of 13-year-olds.

Figure 27: **The percentage of girls aged 15–17 who got pregnant each year, 1975–2012, England and Wales**

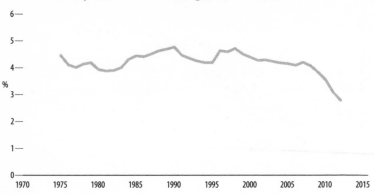

Source: Office for National Statistics.

having sex, which would be fairly remarkable, and is not what we've seen earlier in this chapter. Or are they just being more careful? There have been many attempts at explaining this trend, and it helps to understand not just the overall rate but also its variability around the country, which is extraordinary.

5%: the proportion of 15- to 17-year-old girls getting pregnant each year in Hastings
1%: the proportion of 15- to 17-year-old girls getting pregnant each year in the Derbyshire Dales

Consider the three years 2010 to 2012.[23] In leafy, well-off areas such as Windsor and Maidenhead the annual teenage pregnancy rate was less than 1.5%, while in Middlesbrough and Burnley it was four times higher. Seaside towns have a rather notorious reputation, and Blackpool and Hastings both had rates of around 5% – that's 1 in 20 15- to 17-year-old girls getting pregnant every year, the same as the national rate in 1971. A special study concluded that seaside resorts

Figure 28: **Teenage pregnancy rate compared with child poverty in areas of England and Wales, 2008–10**

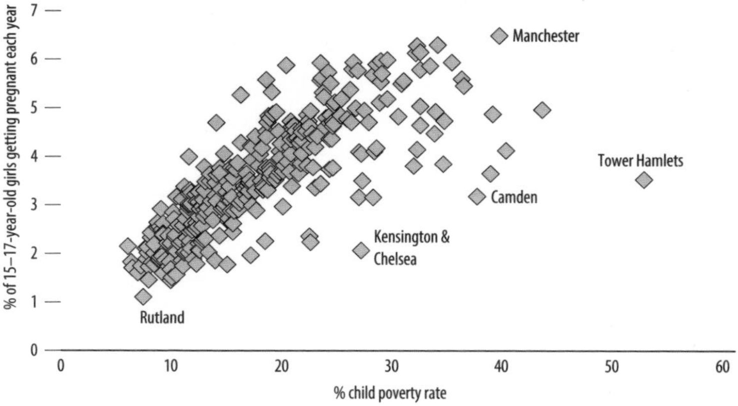

Source: Office for National Statistics.

tend to have a more hedonistic 'carnivalised' atmosphere, with a major entertainment industry and transient populations. This, and easy access to alcohol, encourages unprotected sex.[24]

The strong relation to deprivation can be seen in Figure 28, which looks at the relationship between children living in poverty and teenage pregnancy.[†] The correlation is clear, although there are clear exceptions in which areas have high rates of poverty and yet moderate teenage pregnancy rates: for example, the highest child poverty rates, above 40%, are in the London boroughs of Tower Hamlets, Islington and Hackney, all of which have high ethnic Asian populations with a very different sexual culture. Another interesting place is Kensington and Chelsea, which has a remarkable level of child poverty given its wealthy image, but the teenage pregnancy rate is low, which may reflect the quality of education and services.

[†] The overall correlation is 0.59, and rises to 0.85 in Yorkshire, but is only 0.22 in London.

Teenagers get pregnant because of unprotected, or inadequately protected, sex: the huge drop in the 1970s happened because of better contraception services, but Kaye Wellings argues that the reduction did not continue as it did elsewhere in Europe as services were fragmentary. A joined-up strategy of education and improved access to contraception, combined with a change in aspirations among young girls, is likely to be behind the recent fall.[25]

157: the number of girls who give birth each year, out of every 1,000 aged between 15 and 19, in Malawi
5: the number of girls who give birth each year, out of every 1,000 aged between 15 and 19, in the Netherlands

If you think there's a lot of variation across Britain, have a look at what's going on around the world. The Millennium Development Goals were international targets set by the United Nations in 2000: target 5.B was to 'Achieve, by 2015, universal access to reproductive health', and Indicator 5.5 is the annual birth rate to women aged between 15 and 19: the goal is to reduce such births, whether the mother is married or not, since early childbearing has increased risks both for the mother and for the child.

Figure 29 shows the extraordinary range of birth rates across a selection of countries, representing a wide range of cultures. This single graph deserves an anthropology book just for itself, with societal norms and economic conditions leading to vastly different experiences for young women around the world. For example, the Netherlands has extremely low rates, reflecting later age at first sex and their efficient contraception services. The UK still has a comparatively high rate among developed countries, and the USA and New Zealand similarly combine late marriage with early sexual experience, not all of which is adequately protected,

Figure 29: **Annual birth rate for teenage women for selected countries**

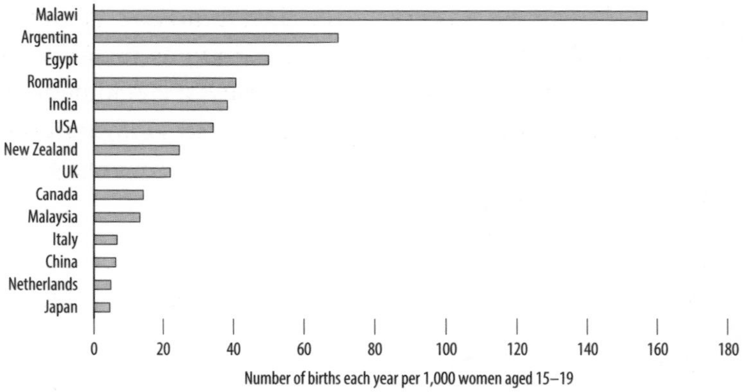

Number of births each year per 1,000 women aged 15–19

Source: United Nations Statistics Division.[26]

although the US teenage birth rate has halved over the last twenty years as use of contraception has improved. Southern Asian countries such as Malaysia have a more traditional culture of sex within marriage, and so a trend towards later marriage has brought a fall in teenage births.

Some cultures that encourage early marriage, such as India and Egypt, have high rates of teenage births, but almost exclusively to married women. Malawi also has a low median age at marriage of 17, but this is combined with early sex outside marriage: one in five girls starts to have sex before she is 15. These factors combine to create a teenage birth rate of 157 per 1,000 – that's one in six women giving birth before 20, over 30 times the rate in the Netherlands. It's difficult to think of any other statistic in this book with such variability. As Kaye Wellings has said, being married does not necessarily protect sexual health: married women may find negotiation over contraception difficult, and very early experience can still be coercive.[27]

We've seen that teenage sexual behaviour shows almost unimaginable variation both between different countries, and across different communities within a country. What young people get up to is influenced by a complex and largely uncontrollable system of social norms, media imagery, economic context and personal relationships, and this defeats simplistic analysis. Nevertheless, behaviour does change, and recently seems to have been changing for the better.

Access to contraception was, of course, vital, but researchers argue that sex education should be far more than just trying to prevent pregnancy and disease. It needs to be about relationships rather than just bodies, and how to negotiate the stages of 'non-coital' activity that young people go through, just as Michael Schofield investigated half a century ago. And teenagers currently negotiating the choppy waters of first love are facing new and complex challenges, dealing with the modern pressures of pornography, sexting and coercion.

It all used to be much simpler. But perhaps less exciting.

8

FEELINGS
ABOUT SEX

In 1973 the historian Carl Degler was looking through papers in the Stanford University archives when he happened to stumble upon a remarkable collection of '650 pages of spidery handwritten questionnaires'.[1]

Degler, who had won the Pulitzer Prize for History the previous year, realised he had made a major discovery, and asked if anyone had ever looked at the files before 'And they said no, no one ever had looked at any of the papers, and certainly not at that survey. That's one of the great experiences of my life as a historian.'[2]

The archive was that of Dr Clelia Duel Mosher (1863–1940), an extraordinary and under-rated contributor to the history of the sex survey. In 1892 she was a biology student in Wisconsin when she was asked to address the Mothers' Club on the subject of marriage, and subsequently designed a questionnaire covering sexual desire, response, contraception and attitudes to sex. She steadily accumulated 45 responses, but she never published the data, and the forms lay undisturbed for over thirty years after her death until their discovery by Degler.†

---

† They have been transcribed into a book – I was the sixth person

Figure 30: **One of Clelia Mosher's reports from 1892**

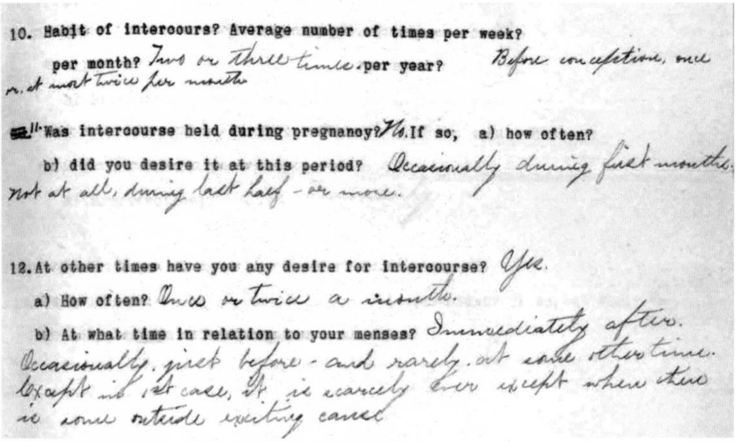

10. Habit of intercourse? Average number of times per week?
per month? Two or three times. per year? Before conception, once or at most twice per month

11. Was intercourse held during pregnancy? No. If so, a) how often?
b) did you desire it at this period? Occasionally during first months, not at all, during last half - or more.

12. At other times have you any desire for intercourse? Yes.
a) How often? Once or twice a month.
b) At what time in relation to your menses? Immediately after. Occasionally, just before - and rarely at some other time. Except in 1st case, it is scarcely ever except when there is some outside exciting cause

From a 25-year-old woman born in 1867 detailing her habit of intercourse of about two or three times a month, and her desire for sex around once or twice a month.[3] Source: Stanford University Library.

Mosher was a doctor and a strong proponent of women's autonomy: she sought to challenge the assumed 'weakness' of females, which she blamed on constricting and heavy clothes, insufficient exercise and the way women were treated. So her questionnaires focused on health, menstruation, pregnancy and so on, but, more importantly, they also explored how women felt about the intimate lives, their attitudes and desire for sex and even their arousal and responses. Her respondents were certainly not a random sample: they appear mainly to have been wives of university faculty, had almost all been to college and worked as teachers before marriage. The majority had been born before 1870, and reported little knowledge of sex before their marriage.

Mosher realised that sexual health was about more than our physical activities. So in this chapter we're going to follow her and look at the statistics of our feelings: our sexual

---

to take this out of the Cambridge University Library in thirty-two years.[4]

Figure 31: **Clelia Mosher**

Clelia Mosher served in the Red Cross in France during the First World War – she was shocked at French attitudes to sex. Source: Stanford University.

desires, whom we feel attracted to, how often we think about sex, how we are aroused and what our attitudes are. These are tricky things to measure, although Mosher boldly tried, and subsequent sex researchers have risen to the challenge.

When do we feel sexual desire?

8: the day of the menstrual cycle associated with peak sexual desire in 110 unmarried women in the 1930s

People researching sexual desire have generally taken for granted that it was an innate characteristic of men, with urges of greater or lesser extent, from mild disturbances to the feeling of being chained to a maniac. Female sexual desire is another matter, and has sometimes been dismissed altogether, in particular by some of the Victorian commentators, such as William Acton, whom we have already met in Chapter 5 pronouncing on the perils of self-abuse:

> The best mothers, wives, and managers of households, know little or nothing of sexual indulgences. Love of home, children and domestic duties are the only passions they feel. [...] She submits to her husband, but only to please him; and, but for the desire of maternity, would far rather be relieved of his attentions.[5]

Acton and his ilk certainly shaped a commonly held perception of the Victorians as buttoned-up, self-denying and sex-avoiding, but this popular 'lie back and think of England' attitude has been disputed.[†]

Degler points out that other voices can be heard, even at the time: popular writers and doctors acknowledged that women had sexual feelings. And Mosher's respondents do not, on the whole, fit the Acton mould. While subject 9, born in 1834, reports that sex is 'usually a nuisance, never much cared for it', she is in the minority. Most consider it at least agreeable, with 35 out of 45 expressing desire independent of their husband: subject 11 (born in 1861) says it is 'very seldom disagreeable, usually very delightful'. Their frequency of sex varies widely, but the median is around three times a month, matching the figure reported by Natsal-3 in 2010. All this rather undermines the popular view of the Victorians,

---

† This archetypal 'Victorian' advice is reputed to be from Lady Hillingdon's 1912 journal 'When I hear his steps outside my door I lie down on my bed, open my legs and think of England', although there is no evidence for this – it certainly does not come from Queen Victoria herself (unless it referred to childbirth), as she and her precious Prince Albert had nine children and appear to have had a very satisfactory sex life.

although we should remember that the attitudes and behaviour of Mosher's American, highly educated group are unlikely to represent what was generally going on in Britain at the time.[6]

It's all very well asking people to count up their sexual activity, but statistics of sexual desire are necessarily based on subjective impressions. Clelia Mosher, perhaps because of her medical background, clearly felt it was associated with the menstrual cycle – perhaps your body would encourage sexual activity at the point where conception is most likely. So, if you are a woman, could you say when you feel 'hot'?

Researching desire is difficult. Given that – as we have seen – the median age for first sex is 16, it would be a challenge in modern Western society to find large groups of young women who would provide data on sexual desire alone, unaffected by subsequent activity. But if we go back seventy years, such studies were possible, although professionally bold given the attitudes of the day. Elsie Widdowson was a young researcher in her late 20s when she and her collaborator Robert McCance began to study how women changed during their menstrual cycle both physically and emotionally.[7] They wanted women, whether sexually active or not, to complete diaries about their sexual feelings, and asked them each day to 'Write 0 if sexual matters are definitely repugnant. Write 1 if sexual feeling is present to a slight degree or easily aroused. Write 2 if sexual feeling is present to an intense degree.' Married women also noted when sex took place.

This was strong stuff for the 1930s, and they recruited with difficulty. When they approached women individually, very few objected to keeping the records, but the problem came when they tried to recruit batches of female students from colleges, whose heads 'took exception to forms on the grounds that sexual feeling was abnormal in unmarried women students and that no forms containing such words

Figure 32: **Pattern of sexual desire over the menstrual cycle**

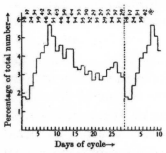

Sum of all the entries from the 21–37-day
cycles = 1246

* These are the actual percentages of
the total number.

**Fig. 11. Variation in sexual feeling
(single women) throughout the
menstrual cycle.**

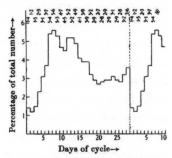

Sum of all the entries from the 21–37-day
cycles = 1618

* These are the actual percentages of
the total number.

**Fig. 12. Variation in the intensity of
sexual feeling (married women)
throughout the menstrual cycle.**

Recorded by 110 single and 57 married women in the 1930s. Source: McCance, Luff and Widdowson.[8]

could possibly be allowed to circulate in their institutions'.

Eventually they managed to get diaries from 167 women, mainly from the educated middle class, who contributed data from 780 cycles. It is a sign of the times that unmarried women were assumed not to be sexually active: 5 out of 115 unmarried women who were actually having sex were included among the 'married'.

Although the 110 'unmarried' women were not having sex, they proved the heads of the colleges wrong by apparently having few problems acknowledging their sexual feelings. Figure 32 shows that the pattern of 'sexual feeling' was similar for single and married women: both show a peak at around day 8, soon after the end of menstruation, and then another more minor peak just before the start of the period.

For the 57 'married' women, sex also had a peak on the 8th day, followed by a sharp drop and then a hump between the 11th and 17th day. The maximum recorded frequency was three times in one day, and twice was 'not uncommon'. The median frequency was four times per cycle – very similar to

142

the rates found in Natsal-3 80 years later – but they noted an impressive maximum of 18 times in one cycle.

Although I would only rank these as 2* data, this was a brave and groundbreaking study for the 1930s: Widdowson had trouble getting it published, and it was twice rejected by the Royal Society.[†10] A recent study, apparently unaware of the research seventy years earlier, wrongly claimed to be the first to compare partnered and non-partnered women, although, unlike Widdowson, they did feel able to ask about masturbation.[11] It took them over seven years to recruit over 1,000 university staff, who submitted to multiple ultrasounds to identify the time of ovulation. They found that there was more sexual activity around ovulation, particularly in single women, but sexual activity in partnered women depended far more on whether it was the weekend or not.

There's been a mass of research showing that individual women do report variation in desire over their cycle, but after all this time no consistent pattern has been determined, which perhaps is unsurprising given that mundane social arrangements appear to dominate biological impulses.[12] Once again, the extraordinary variety of human experience is not easily summarised by a few statistics.

---

†Widdowson and McCance became a legendary pair of researchers who in 1940 produced 'The Composition of Foods', known as the 'bible for dietitians' – it was still in print over sixty years later.[9] This formed the basis for the rationing system of the Second World War: they tried the rations out on themselves, and found they could survive well on little meat or dairy products, provided they took calcium supplements. Elsie Widdowson lived in a village outside Cambridge looking after her mother, who lived until she was 107: Widdowson herself died in 2000, aged 93.

How often do we think about sex?

~~~~~~

10: the number of times a day that an average female student
thought about sex (2*)
15: the number of times a day that an average female student
thought about food (2*)

~~~~~~

Another great statistical 'fact' is that men think about sex every seven seconds, although I have failed to find the origin of this 0* statistic. But how often people think about sex is clearly such a compelling question that, despite the obvious challenges, researchers have tackled it. As always, one option is just to ask people: in the 1992 US National Health and Social Life Survey (NHSLS), 54% of men and 19% of women said they thought about sex 'every day' or 'several times a day', which hardly shows obsession.[13] An alternative is to get people to try to record when they think about sex: in one study 283 college psychology students carried around golf-clickers for a week, recording when they thought about sex, food and sleep, with the instruction 'If you think about what someone to whom you are attracted might look like naked, you will add a tally with the tally counter.'[14] So presumably if you heard a lot of clicking in a lecture, you knew what people nearby were thinking.

Men estimated in advance that they thought about sex 5 times a day, whereas actually they clicked a median of 19 times a day: women predicted they thought about sex 3 times a day, but actually clicked 10 times on average, so it seems people are not very good at judging their mental interest. The maximum for a man was 388 recorded thoughts in a day, and 140 for a woman – they must have been clicking at everyone they saw, possibly including on TV. But men thought about food almost as much as sex, a median of 18 times, and women thought about food 15 times a day, rather *more* than they thought about sex. Even though these must

144

be 2* numbers, at best, on balance they suggest there was not much difference between men and women, and food was, essentially, just as important as sex.

But can you really focus all day on recording your thoughts, and would that, in fact, start to shape what you think about? German students were given BlackBerry smartphones for a week, and sent seven signals a day at random times asking what desires they were either feeling currently or had been feeling in the last half-hour.[15] Try thinking what it would be like in such an experiment: if a random call came at you now, could you choose your current desire from a list? They added a financial incentive to respond, ensuring a 92% response rate. Eating, drinking and sleeping came well above sex in how often the desire was felt, although desires for sleep and sex were the strongest.

So sexual desire is important but not, on average, the mythical constant distraction to men (except to the gentleman who clocked up 388 thoughts a day).

Psychologists, of course, have developed scales for measuring sexual desire in more detail. The Sexual Desire Inventory (SDI-2) has been popular,[16] which asks, for example 'During the last month, how often have you had sexual thoughts involving a partner? From (0) not at all, to (7) many times a day' and 'When you first see an attractive person, how strong is your sexual desire [on a scale 0 (no desire) to 8 (strong desire)].' If you want to check your levels, you can do so online, although be warned that any score below 45 will be classified as 'low desire'.[†]

---

† The form can be accessed at www.centerforfemalesexuality.com/sexual-desire-inventory.php. It makes a very simple division: any score of 46 or more produces the response 'Your sexual desire is at a healthy level', whereas any score of 45 or less produces the warning 'You may be suffering from low desire and may be helped by contacting a practitioner who specialises in the medical diagnosis and treatment of female sexual dysfunction.'

Who do we find attractive?

0.71: the waist:hip ratio of Jessica Alba (3*)

In the 1880s the eminent statistician and eugenicist Sir Francis Galton[†] wanted to create a Beauty Map of the UK:

> I use a needle mounted as a pricker, wherewith to prick holes, unseen, in a piece of paper. I used this plan for my beauty data, classifying the girls I passed in streets or elsewhere as attractive, indifferent, or repellent. [...] I found London to rank highest for beauty: Aberdeen lowest.[17]

One wonders what the young women of Aberdeen thought about this eccentric gentleman, staring at them and fiddling in his pockets; perhaps they reciprocated his feelings.

There are numerous online lists of what characteristics people find sexually attractive,[18] but how do we measure the attractiveness of an individual objectively? The main method used in sex research is 'self-report', in which we ask people what they find attractive, to rate characteristics or rate a hypothetical partner. The same BBC Survey that tried to measure digit ratios (see Chapter 4) also asked 'What features do you consider most important in a partner?', and 200,000 volunteers were allowed to pick three. Overall the

---

†Francis Galton (1822–1911), a cousin of Charles Darwin, was an eccentric polymath who was obsessed by measurement. He invented weather maps and the idea of using fingerprints in criminal cases, as well as developing statistical ideas such as regression, correlation, regression-to-the-mean and so on. He also invented the term 'eugenics' for the movement that wanted to improve the 'race' by selectively encouraging the 'fit' to have more children and the 'unfit' to curtail their breeding. People such as Galton would decide who was 'fit'. He had no children.

most important traits were, in order, intelligence, humour, honesty, kindness, overall good looks, face attractiveness, values, communication skills and dependability, although men ranked good looks and facial attractiveness higher than did women, while women ranked honesty, humour, kindness and dependability as more important than did men.

Although these findings are hardly surprising, they do force us to consider what a 'partner' is: what people find desirable in a long-term partner may be very different from what they may want from a more temporary relationship, maybe even a one-night stand, where the emphasis is on sexual attraction.

One study asked 320 Australian students a series of questions about the characteristics that would make good prospective partners, but how these changed if the aim was only an intimate relationship.[19] When it came to long-term partners, three factors were identified that best explained the overall pattern of answers; they termed these 'warmth/trustworthiness' (understanding, supportive, considerate, kind), 'vitality/attractiveness' (adventurous, nice body, outgoing, sexy) and 'status/resources' (good job, financially secure).

But for those merely seeking intimacy rather than an established partner, the important factors were rather different: 'intimacy/loyalty' (honesty, commitment) and 'passion' (exciting, challenging). While this study has hundreds of citations in the academic literature, it is worth remembering that the average age of the participants was 22. We can't expect our interests to remain static from our early twenties, even if their conclusions do seem fairly reasonable and perhaps rather predictable: ideal partners are nice, fun and secure, while ideal relationships are honest and exciting.

Personality is genuinely important, but physical attributes doubtless play a part in attraction, at least in the early stages. Research into bodily attraction has concentrated, somewhat predictably, on the geometry of the female form.

147

In a classic 1993 analysis of the (reported) measurements of *Playboy* centrefolds from 1955 to 1990 and Miss America contestants from 1940 to 1985,[20] it was found that the ratio of waist to hip measurement had stayed almost constant at 0.7, in spite of models getting slimmer: 0.7 could mean, for example, a waist of 28 inches (71 cm) and hips of 40 inches (102 cm). When shown line-drawings of women with different waist:hip ratios, 0.7 was chosen as the most attractive by both men and women, although initially 0.7 was the smallest ratio offered: when a wider range of computer-manipulated images was used, attractiveness still peaked at 0.7 for both male and female judges.[21]

So Marilyn Monroe, who was reported to have a waist:hip ratio of 0.64,[22] might be considered rather too hourglass compared with Jessica Alba's 0.71.[23] But there is reasonably consistent evidence that, on average, waist:hip ratio is more important than breast size when it comes to simple bodily attraction to men, at least in our Western society. This has been claimed to be an evolutionary attraction to good health and fertility, but it's more plausibly a result of our current cultural norms.

10: the relative importance of education compared to genes when it comes to choosing a partner (3*)

Instead of relying on people's opinions about whom they find attractive, we could cut to the chase and look at who ends up with whom. The 1992 US National Health and Social Life Survey (NHSLS) showed that people in the US tended to marry someone who was very similar to themselves, with similar ethnicity in 93%, similar age (78% less than five years' different), similar education (82% no more than one category difference), and 72% with the same religion.[24] A recent study of the DNA of 825 US couples has shown that we tend to choose spouses that are genetically similar, but this is nowhere near as important a factor as education.[25]

Isn't it amazing that the 'only one in the world' happens to be so similar? It must have meant to be! But it's hardly surprising that partners have a lot in common – whom we meet and fall for is determined by our family, social and work life, and we are simply physically more likely to come across someone who is like us. Then we also feel more confident in initiating a sexual relationship with someone similar, not least through having the approval of our social network. All those fine features the Australian students identified that make individuals attractive seem plausible, but maybe they could have simply said 'similar to me'.

Measuring arousal

93%: the proportion of women who, after filling in a sex history, refused to take part in laboratory measurement of their sexual arousal

If we want to know about the physiology of sex, about what actually happens to bodies when they are aroused and active, we enter the domain characterised by the work of Masters and Johnson: the sex laboratory.† But do any statistics that come out of such a laboratory apply more generally? We have seen that people who volunteer to answer questionnaires about their sexual activities may not be entirely representative. And, if anything, this applies even more to the sort of personalities who agree to take part in laboratory work – to be wired up, constantly observed, filmed.

In a well-designed study, 484 psychology students (of

† See *Bonk* for an entertaining romp through laboratory sex research, with hardly a statistic in sight.[26]

course) at two US universities completed a sex-history questionnaire and were then asked if they wanted to proceed to a
'sexual arousal' study, 'during which you will be required to
watch erotic videos depicting explicit sexual scenes'.[27] This
may not have seemed a bad way to earn research credits,
but the men were then told that 'the size (circumference) of
your penis will be measured by a strain gauge device. This is
simply a small flexible piece of rubber tubing which you will
place around your penis after the experimenter has left the
room', while the women were offered the following enticing
opportunity: 'the amount of blood flow in your vagina will
be measured by a photoplethysmograph gauge device. This
is simply a small plastic tampon-shaped instrument which
you will insert into your vagina after the experimenter has
left the room.' Only a small proportion volunteered; 53 of the
198 men (27%) and 21 of the 287 women (7%).

The main reason for refusal, perhaps unsurprisingly, was:
'I don't think I would be comfortable having my sexual
arousal measured.' The 74 who did volunteer differed systematically from the non-volunteers in having experienced
more sexual behaviours, more previous partners and less
sexual guilt. As the researchers ruefully acknowledged, 'This
points to an urgent need to develop less intrusive measures
of sexual arousal which may attract more representative
samples of participants.' Maybe the setting could be made
more intimate: one small study got women to watch erotic
videos both at home and in the laboratory, and reported that
physiological arousal was stronger in the comfort of their
own home.[28]

Despite a huge range of devices being developed to
measure physiological arousal in women, ranging from the
photoplethysmograph mentioned above to the 'electrovagi-
nogram' (best not to ask), a recent review bemoaned the lack
of correlation between what was measured by the devices
and what women actually said they were feeling.[29] In general,

laboratory studies may be fine to show what *can* happen but are unreliable guides to what *does* happen in general, which is why they hardly feature in this book. But there have been experiments on the effects of sexual arousal that are just too interesting to ignore.

The effect of arousal

'6': the level of disgust that female students associated with putting their hand in a bowl of apparently used condoms, having been primed with a pornographic film

We saw in Chapter 4 how 'Gerald' found the idea of oral sex 'disgusting', and when we think of the normal aversion to other people's genitals, bodily odours and fluids such as saliva, sweat and semen, it's perhaps a wonder that anybody can stand to have sex at all. But we do, and this suggests that, when we're sexually aroused, these standard responses seem to change, and, of course, some imaginative researchers have worked out how to test this.

They randomly allocated 90 young female students to watch one of three 35-minute films intended to influence their mood: 30 saw female-friendly erotica (the 'sexual arousal' group), 30 saw a film of exciting activities such as sky-diving and mountaineering (the 'positive arousal' group), while the remaining 30 were left with watching a film of a train ride (the 'neutral' group, although for a train-nerd like me this would probably have counted as positive arousal).[30]

They were then all invited to participate in sixteen 'disgusting' tasks with both non-sexual and sexual associations, such as putting a needle in a cow's eye ('a new eyeball was brought every day from the butcher and taken back at the end of the day for proper biohazard waste removal') or touching each of a bowl of 'used' condoms (actually covered in lubricant).[†] The 'sexual arousal' group found all the tasks less disgusting than the other two groups, particularly the ones with a sexual connotation: for example, the condom bowl was on average rated as '6' on a 'disgusting' scale from 0 to 10 by the sexual arousal group, and '7' by the others, while the cow's eye topped the disgusting league table.

I am not sure what lesson to take from this: maybe next time the dog throws up, watch some pornography before cleaning it up?

Attitudes

21%: the proportion of 65- to 74-year-old British men who agree that male same-sex relationships are 'not wrong at all' (3*)

Soon after the Kinsey report appeared in 1948, a little-known study nicknamed 'Little Kinsey' was carried out in Britain by Mass Observation (MO).[31] MO was a well-known social research organisation that examined the British public much as anthropologists used to investigate remote tribes, and for this exercise they sent panels of 'observers' into dance halls and pubs to report on conversations and behaviour. But the *Sunday Pictorial* newspaper also paid for a 'random' sample based on stopping over 2,000 people in the street, standard practice for surveys in the late 1940s. There was also a postal survey of 1,000 of each of three groups of 'opinion leaders': clergymen, teachers and doctors. Mass Observation also had a fixed National Panel, much like a modern online panel, but

---

†Other activities included taking a bite from a biscuit next to a (real) large live worm, holding a bandage used on a wound (faked), reading a brief pornographic story and saying out loud 'It was so horny to have the dog inside me', placing used women's knickers (actually spoiled with drops of coconut milk) in a bag and so on. The experimenters clearly had vivid imaginations.

whose membership 'consists predominantly of middle class people of more than average intelligence'.

There is dramatic variation in opinions between these different groups. When asked if they disapproved of sex outside marriage, only 24% of MO's own panel said yes, compared with 63% of the street sample, 65% of doctors, 75% of teachers and, predictably, 90% of clergymen. And 83% of the MO panel approved of divorce, compared with 57% of the street sample and 33% of clergymen. As well as liberal views, MO's panel also had some remarkably liberal sexual behaviour.[†]

The National Panel also fed back vivid descriptions of their sexual experiences, many reflecting dissatisfaction among women with male lack of intimacy.[‡] 'Little Kinsey' was not properly published at the time and has remained a largely undiscovered gem: possibly the first attempt at a sex survey partly based on a 'random' sample.[32]

Fortunately, we now have a great opportunity to study changes in attitudes over recent history, as Natsal has asked the same questions in successive surveys – these concern extramarital sex, one-night stands and attitudes to same-sex partnerships.

Figure 33 shows there has been a substantial shift in attitudes between Natsal-1 in 1990 and Natsal-3 in 2010. In the top left we can see that both men and women have become increasingly intolerant of marital infidelity, with more women believing it is always wrong. Top right, it is clear that a fairly steady 25% of men have a tolerance of one-night

---

[†] 400 of the MS's National Panel returned questionnaires. 25% of the men had been with a prostitute, two-thirds of them within the previous five years. 38% of married people had had sex before marriage, and 24% said that they had had sex with someone else during their marriage. 12% had had physical homosexual relationships, and 8% had homosexual 'leanings'.

[‡] For example, 'my husband accused me of being "cold" but little knew the passionate longing I experienced. If only he had made love to me instead of using me like a chamber pot.'

Figure 33: **Percentage of 16–44-year-olds in Britain agreeing with each statement in 1990, 2000 and 2010**

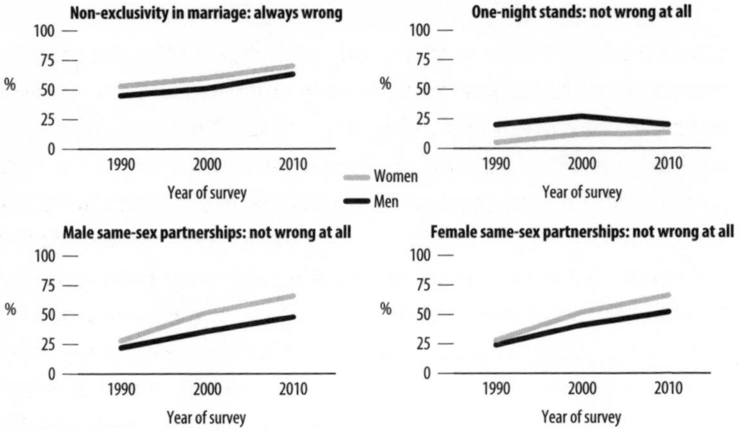

Source: Natsal surveys.

stands, while fewer women believe they are acceptable, but the gap between men and women is narrowing.

When it comes to same-sex partnerships, these are now acceptable to the majority of men and women, up from just 25% in 1990, with women consistently finding them more acceptable than men. Although, as the number at the start of this section indicates, older generations may be slower to change their views.

Of course, all these crude percentages cannot reflect the wide range of views held by individuals or communities within Britain, from libertarian to traditional conservative behaviour. When we take a global perspective, we can see these different perspectives reflected in whole countries – in the Pew Global Attitudes Project, less than 15% of people in Spain, France and Germany thought homosexuality was 'morally unacceptable', compared with over 95% in countries such as Ghana, Egypt and Jordan.[133] Although when it

---

† The Pew Research Center's 2013 Global Attitudes survey is based on 40,117 respondents in 40 countries, from a mixture of face-to-face

comes to having affairs, the only country in which less than half thought it morally unacceptable was ... France.

Clelia Mosher also asked her sample of 45 Victorian women about their attitudes, particularly on the purpose of sex. Thirty considered 'reproduction' as the primary purpose, which meant that a third thought there was an even more important reason; for example, one women, born in 1860, said: 'We believe in intercourse for its own sake – we wish it for ourselves and spiritually miss it, rather than physically, when it does not occur, because it is the highest, most sacred expression of our oneness.' It would be interesting to know what a modern sample would consider the purpose of sex.

Mosher spent her life trying to improve the physical health of women, but she had strong eugenic opinions, and her lectures to women in the 1920s emphasised their obligations to marry and have children in order to improve the stock: she would probably be appalled at some of the changes that have taken place. And underneath her 'mannish clothes and shapeless tweed hat, unchanged in style for thirty years' lay hidden depths: an unpublished romantic novel was found in her papers. Kara Platoni has written movingly that 'ultimately, Mosher's story is deeply ironic: She was a staunch feminist who remained aloof from sisterhood, a woman who rigorously researched sexuality and marriage yet probably experienced neither, a pioneering scholar who longed for recognition but did not live to enjoy it.'[34]

Mosher achieved her ambition to be independent and have a career, becoming a full professor at Stanford in the year before she retired. Then, like Elsie Widdowson, she looked after her ageing mother and carefully tended her immaculate garden.

---

or telephone interviews, carried out by a market research company. A random design is used: refusal rates are not quoted, but attitude surveys can get good response rates. I think these are 3* numbers.

9

TOGETHER AT LAST: BECOMING A COUPLE

Sex, mostly, involves two people. And most of the sex that goes on is between two people who have formed a 'couple' – who are seen by themselves and others as being in a long-term relationship, whether married or not. We've seen how sexual activity tends to decline with the length of time together, but now we're going to look at how it all starts, starting with how many people have had sex before getting married, a tell-tale sign of which is being pregnant on the wedding day. Such is the power of sex and statistics, we can even delve into the sex lives of young couples up to 400 years ago, when, for most, marriage signalled the start of sex. We'll see the dramatic change in partnerships and sex: not only over 400 years, but over the last generation. And we'll finish with the great mystery of the disappearing Victorian babies.

Sex before marriage

97%: the proportion of survey respondents in Indonesia who agree that sex between unmarried adults is 'morally unacceptable'

Nowadays the obsessive concern of previous generations

about sex before marriage seems almost as anachronistic as Bakelite telephones and cars with crank handles. And even when social pressure is exerted on young people to keep them celibate, as by Christian campaigning groups in America – we've seen that few young people now wait to get married before they have sex, no matter how enthusiastically they pledged. Of course, it hasn't always been this way, and those who were caught out – usually by the arrival of a baby – often paid a heavy price.

However, a somewhat more lenient approach has traditionally been taken to couples who were formally or informally betrothed to each other. A few centuries ago in Europe and North America this might have included the practice of 'bundling', in which a couple could spend the night together provided they remained dressed, or stayed in separate sacks. When we come to the 1930s, Szreter and Fisher report numerous stories of sex starting during an engagement: 'Agatha', born in 1910, was a domestic servant engaged to a gardener and felt it was 'above board' to have sex, as 'it wouldn't have made any difference if I'd got into trouble'.[1] Even by Kinsey's time in the 1940s, pre-marital sex really meant what it said: he estimated that 50% of women were virgins when they married, and about half of the others had had sex only with their fiancé. Of these, 41% had had sex in cars. Rates rose steadily: of British women married in the late 1950s, 35% had had sex with their husband before marriage, while among those marrying in the early 1970s it rose to 74%.[2]

These changes in behaviour reflect the rapid alteration in Western attitudes to pre-marital sex: in the Global Attitudes Project introduced in the last chapter, less than 10% of people in Spain, France and Germany thought sex between unmarried adults was 'morally unacceptable', compared with over 90% in countries such as Pakistan, Jordan or Indonesia.[3] Over the forty countries surveyed, the median proportion who thought it morally unacceptable was 46%, which, given the huge diversity across different cultures, is possibly one of the most useless averages I can think of.

It's easy to ask about attitudes, but it's more challenging to find out how many people have sex before marriage now, and did so in the past. But there is a neat way to get at least a lower bound on this, without actually asking anybody to give away any secrets. That's to look at the number of women who were pregnant when they got married. And we'll find it is quite a lot.

'Pre-nuptial pregnancy'

38%: in 1830, the percentage of brides who were pregnant when they married (2*)

Throughout history parents have noticed a suspicious bulge, a tearful confession has been wrung out, and hasty arrangements for a marriage are made … but how often has this actually occurred? For a statistician this is a fine example of where some imaginative use of official statistics can tell us something about secret human behaviour. The basic idea is, in principle, straightforward: combine marriage and birth records and see how many births occurred within eight or so months of the wedding – each must be the consequence of what is known as a 'pre-nuptial pregnancy'. Of course, this provides only a very rough lower bound of the proportion having sex before marriage: many will have done so and managed to get away with it, leaving nothing for future statisticians to count.

After 1837, the year an 18-year-old Victoria came to the throne, we can get data from the British national civil registration system: this would be perhaps 3* initially and 4* when the records later became reliable. Before this, a system of parish registration had operated since 1538, when it was started by Thomas Cromwell on behalf of Henry VIII: long lists of baptisms, marriages and funerals, written with a quill pen and kept for centuries by successive clergymen.

Yet mining this rich but dusty seam of information is an enormous undertaking. The data I use in this book come from the Cambridge Group for the History of Population and Social Structure, which in 1964 amassed an army of volunteers to collect summary data from 404 parish registers covering England between 1580 and 1830. All this information had to be punched into primitive computers, and finally, after a mere twenty-five years' work, the conclusions were published, using clever statistical methods to estimate the size of populations, the birth and death rates, average lifespan and so on.[4]

But if we want to look into pre-nuptial pregnancy, we need to connect marriage to birth records, and so need full family reconstitution,[†] which involves a process of linking records across generations in a way familiar to anyone who uses modern genealogy software. So twenty-six parishes with the best records were selected, ranging from Earsdon in Northumberland to Hartland in north Devon.[5] These provided some remarkable data.[‡]

Figure 34 shows the long but changing tradition of brides being pregnant on their wedding day: the rate was nearly 30% when the 'virgin' Queen Elizabeth ruled and Shakespeare was writing plays of teenage love, then declined to around 18% during Oliver Cromwell's rule in the 1650s, as

---

[†] Not to be confused with *family reconstruction*, which is a form of therapy.

[‡] The family reconstitution data have been criticised for using unrepresentative parishes, ignoring migrants between parishes and missing nonconformists, and for their limited ability to link records.[6] Clearly the parish records are not a 4* source, and the statistics may only apply to Anglicans in rural areas: but I would consider the results as at the high end of 2* data, and still an amazing achievement.

Figure 34: **The percentage of marriages in which the bride was pregnant between 1580 and 1830**

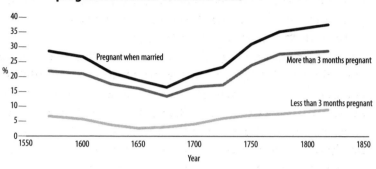

Based on a total of 27,000 marriages: about 100 a year for 250 years. The lower lines break the total into those who were more or less than three months pregnant: if less than three months, the marriage occurred soon after the pregnancy was confirmed.[7]

the Puritans, who took a decidedly dim view of fornication, came to power. The rate then increased steadily through the decidedly non-Puritanical 1700s until reaching nearly 40% when Victoria was crowned.

It would be wrong to see all these individual stories in terms of the hackneyed 'shotgun wedding' story. In some cases the baby was not born until they had been married more than seven months, so the bride had conceived at most two months before the wedding. The graph shows that these formed a remarkably regular 10% of all marriages: the couple would probably have already been betrothed with the agreement of the families, a wedding was being planned, and they just jumped the gun a bit.

The other type of pre-nuptial pregnancy has the couple baptising the baby within six months of getting married, so the bride was more than three months pregnant: in at least some of these cases we might presume the couple were not betrothed, the pregnancy was unplanned and the bulge is visible as the bride goes down the aisle amid wagging tongues.

There could be an additional complication, even when the pregnancy was just a reason to bring the wedding date forward a month or two.[8] Over much of this period the Ecclesiastical Courts still had some jurisdiction, and considered what was bizarrely known as 'fornication before marriage with his own wife' as an – admittedly minor – crime. These 'bawdy' courts dealt with sexual matters up to the end of the 1700s, although the punishment for this offence was usually a penance of appearing in church and publicly confessing, which may have been only a minor embarrassment since the situation by then would have been common knowledge.

If you were a bit older, you were more likely to have things planned: over half the pregnant brides over 35 were in the early stages. Rather predictably, the pattern is different for teenagers: only a quarter of pregnant teenage brides got married immediately they knew they were expecting: one can only imagine the lengthy and heated family discussions that went on before the other three-quarters finally tied the knot.

It is tempting to view the high rates of pre-nuptial pregnancy in the early 1800s, which presumably accompanied more pre-marital sex, as a sign of a more liberal and relaxed approach to sexual expression. But we've already seen that this was a time of increased condemnation of masturbation and sexual activity in general. So how can we resolve this apparent paradox? It's been suggested that the repression of any form of sexuality other than standard unprotected penetrative intercourse, and an increasing emphasis on female passivity, led to a 'phallocentric' focus in which men increasingly felt it was necessary to have 'proper' sex – the value of alternative means of stimulation was downgraded.[9] Rather what Shere Hite was campaigning against two centuries later.

51%: in 1938, the proportion of brides under 20 who were pregnant when they married (4*)

Figure 35: **The percentage of brides who were pregnant when they got married, 1951–2006, England and Wales**

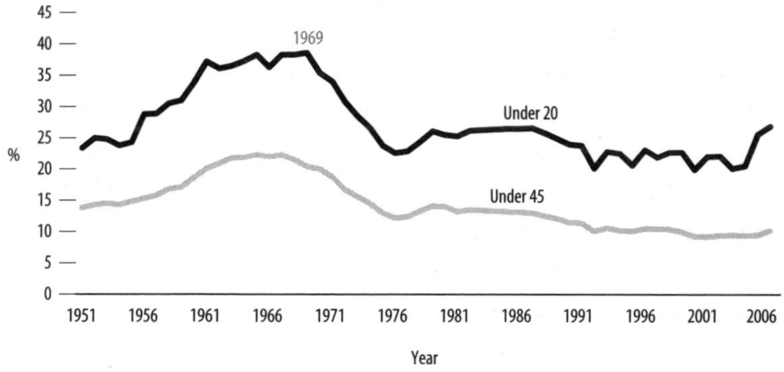

Source: Office for National Statistics.

A further century of propaganda against pre-marital sex had limited impact: in 1938 the Registrar-General found that 18% of all brides in England and Wales were pregnant on their wedding day, and the rate went up to 51% for brides under 20. The Registrar-General was clearly shocked, having the (rather obvious) 'strong impression that a large proportion of our children continue to arrive other than at the conscious and deliberate intention of their progenitors'.[10] Each case represented an individual drama that is not lessened by being repeated by the thousand: 'Mavis', born in 1908, told Szreter and Fisher that she was unlucky as they had 'only had it once', but her father 'went mad, he did', and she had to get married immediately. Fortunately her father eventually calmed down, and ended up thinking her husband was one of the best in the world.[11]

Figure 35 shows what happened to pre-nuptial pregnancy after the Second World War, based on official 4* data. Rates in the early 1950s were lower than just before the war, with around one in seven brides pregnant. By 1965, with increased sexual activity not being matched by readily available contraception, this had risen to 22% of all brides and nearly 40%

163

of brides under 20. This is not quite as high as the rates in the 1830s, but still meant a lot of mandated marriages.

With the advent of effective contraception these rates almost halved in the 1970s, and by the time the Office for National Statistics stopped publishing these calculations in 2006, around 10% of brides were pregnant when married, probably the lowest rates that have ever existed in this country. And in most of these cases the pregnancy probably simply provided the impetus for long-term, committed partners finally to tie the knot. This fine statistic had lost its remarkable power to reveal transgressive behaviour.

Illegitimacy then

10%: the proportion of babies born in Norfolk in the 1840s who were illegitimate (3*)

For those who didn't have anyone to take them down the aisle, shame, and possibly social isolation, awaited. But unlike other disapproved sexual behaviours – such as incest or masturbation – babies born 'out of wedlock' can be measured with official statistics.

Figure 36 pieces together data from the family reconstitution studies with official statistics since 1837 to show the historical record in England between 1585 and 1960, which is just before the 'illegitimacy ratio' – the proportion of babies that were illegitimate – started to take off.† Over this period

---

†We could look at the illegitimacy *rate* (i.e., the proportion of women who have illegitimate babies each year) or the *ratio* (i.e., the proportion of babies who are illegitimate). We will use the ratio, which is easier to interpret, as it is not so influenced by the general fertility of population. The ratios before 1837 are obtained from baptism records in those few parish registers that recorded illegitimacy, which are then scaled up by 20% to allow

Figure 36: **Percentage of all births that were illegitimate, England 1585–1960**

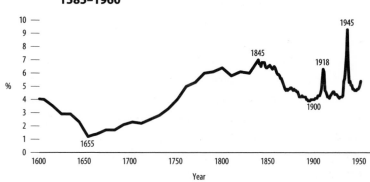

Sources: 1580–1837, from reconstitution from parish records,[12] and 1846–1960, from official registrations[13]

the ratio wobbled between 1 and 8% of all births, and a little knowledge of English history helps in interpreting this graph. As with the proportion of brides who were pregnant at their wedding, the lowest time is during the Puritan rule, between the execution of Charles I in 1649 and the restoration of his son as Charles II in 1660. But this does not necessarily mean that Oliver Cromwell managed to prevent illicit sex; registrations may have been avoided, or couples may just have been a bit more careful.

Then there is a steady increase over the 'long eighteenth century' between the 1680s and the 1830s, with a peak ratio around 1850, when the Registrar-General published league tables of counties that allowed earnest and moralising statisticians to produce maps of 'bastardy' data. These were some of the first infographics. Figure 37 comes from Fletcher in 1849, who considered that 'an excess of bastardy is a fair test of the extent of rude incontinence prevailing among the population at large'.[16]

---

for unregistered babies.[14] The same names are reported to crop up repeatedly in the parish registers: Laslett refers to them as a 'shadowy society of bastard bearers'.[15]

Figure 37: **Bastardy map of England and Wales from 1842**

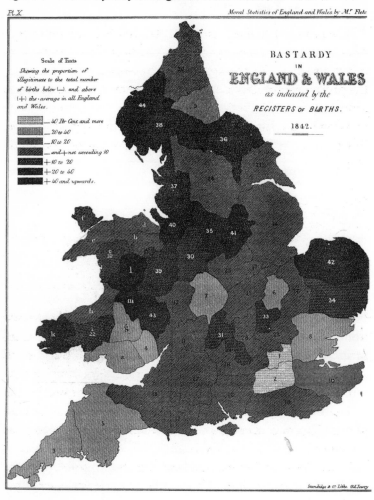

Note the darkness of Norfolk, indicating high rates. Source: Statistical Society of London.

Living in East Anglia, it does not escape me that Norfolk just about tops the bastardy league, with a ratio of 9.9%, beaten only by Cumberland at 11.4% and Hereford at 10.6% – a fair amount of 'rude incontinence'. The figures for the poorer parts of London are, however, suspiciously low, such as Poplar declaring 2%, and the Registrar-General admits that this is probably due to illegitimate births not being registered.

The ratio fell as the Victorians steadily became more continent, and then from 1900 to 1960, apart from bursts during the two world wars, when social strictures weakened, the proportion of illegitimate births remained fairly stable at around 4 to 5% – about one in twenty births being outside marriage – the same ratio as existed around 1770 and 1870. But then the Sixties started, and the rate started its steady climb.

Illegitimacy now

52%: in 2012, the proportion of babies born in England and Wales whose parents were married (4*)

Figure 38 reveals a massive change in social norms, as the number of births outside marriage rose to almost half of all births, in just fifty years.[17] Note that the scale for Figure 36 only goes up to 10%, while that of Figure 38 goes up to 50% – if the graphs were on the same scale, this would be the ultimate 'hockey stick' (although any relationship to atmospheric CO_2 concentrations is presumably correlation and not causation).

If we look at births that are registered solely by the mother (the lower line), we see this has stayed around 5% of all births, essentially the same rate that has operated for century after century. The growth has been in children registered to both parents and so presumably in a reasonably stable relationship. From 1986 we have data on whether the co-registered parents shared the same address, and these cohabitants form the majority of unmarried parents. So in 2012, out of 100 'illegitimate' children, 12 were registered solely in their mother's name, 22 had two registered parents

Figure 38: **The percentage of births that were illegitimate, England and Wales, 1960–2012**

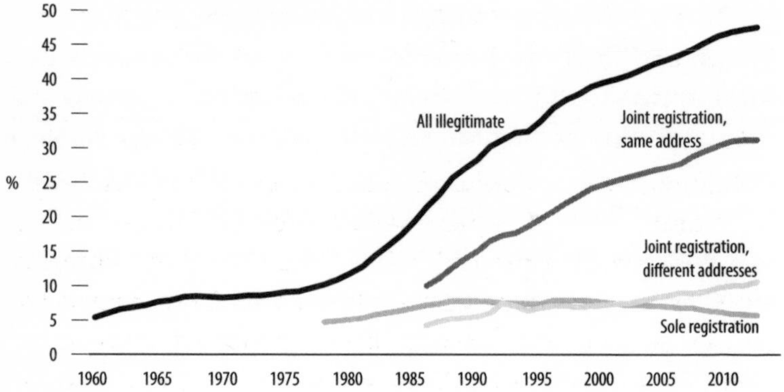

Source: Office for National Statistics.

but at different addresses, while 66 had two parents living as a family unit at the same address.

In 2012 the overall proportion of births in England and Wales to an unmarried mother reached 48%, but this ratio depends very much on where the mother was born. Around three out of five children born to mothers born in Central Africa or the Caribbean do not have married parents, compared with less than one in ten whose mother was born in Bangladesh or India.

These cultural differences become even more stark when we look at international data. Many Western countries have seen an even greater separation of birth and marriage than in Britain, with more than half of births being 'out of wedlock' in France, Sweden and Norway, and with Iceland topping the league table at 63% born outside marriage in 2010.[18] In southern Europe there has been less change to the traditional pattern, with only 6% of births in Greece being technically illegitimate. And in Japan and Korea things have not changed at all: couples still wait to be married before having children and so the illegitimacy rates are less than 2%, a rate that only the fear of Oliver Cromwell could create in Britain.

When do people get together?

〰〰〰〰〰〰〰〰〰〰〰〰〰〰〰〰〰〰〰〰〰〰〰

10%: in 2005, the proportion of cohabiting
25-year-olds who were formally married

〰〰〰〰〰〰〰〰〰〰〰〰〰〰〰〰〰〰〰〰〰〰〰

We know quite a lot about what went on in past centuries. If we want details of important dynasties, buildings, battles, laws and even the price of wheat, then there are libraries full of documents, repeatedly analysed and interpreted by historians. But before the time of sex surveys, there is a pervasive silence about sexual behaviour – we may know about the arranged marriages of the nobility, but neither men nor women left traces of how they negotiated their subsequent activity, with its perpetual attendant risk of pregnancy. All we've got to go on are records of the consequences – the births of children, and all too often the deaths of the mothers.

But we know that the primary determinant of the amount of sex going on is cohabiting relationships, now and even more so in the past, when sex outside marriage carried such social disapproval. So if we assume that sex started soon after marriage – possibly very soon indeed – the age at which people got married can give us a rough upper bound on when sex started.

Figure 39 shows the average age when men and women got married in Britain between 1610 and the present. Unlike traditional Mediterranean cultures, where a new wife would essentially move in with the husband's family, in Britain there is a long tradition of a new couple setting up home independently. So when the economy was going well and there was some work and some money, it was easier for a couple to set up, and they married younger – it's not just a matter of meeting the 'right' person.

In the 1600s people married late, and around 20–25% of the population did not marry at all, but then as the economy

Figure 39: **The average age at marriage for males and females in England, 1610–2010**

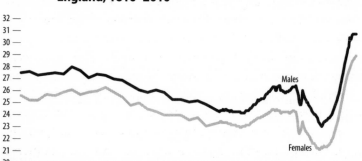

Sources: the statistics between 1610 and 1830 are based on family reconstitution studies and represent the mean age. From 1846 the data come from the Office for National Statistics and the median is given, and just for first marriages, which better represents the experience of the average person (although in this situation the mean and median will be close, as there is a very limited 'long tail' of people getting married when they are very old).[19]

improved in the 1700s, and Britain went into the industrial and agricultural revolution, the average age at which women married fell from around 26 in the 1710s to 23 in the 1870s – this may not seem a lot, but this added another three years of childbearing, and the population rose steadily.

If we look back at the illegitimacy graph on p. 165, we see that during the 1700s the illegitimacy rate went *up* as the people started marrying *younger*. A naïve expectation might be that illegitimacy would go down, as there would be less time to have pre-marital sex. But the opposite happened, suggesting there isn't some biologically fixed amount of sexual expression which must come out either inside or outside marriage: the historical context seems to be the main determinant of behaviour, and there was simply a lot more penetrative sex going on in the 1700s.

Between 1870 and 1930 people started to put off marriages until they were older, and then there is an eloquent dip in 1943 and 1944: couples got married young amid uncertainty about the husband's future, sadly reflected in the number of married men in the Second World War cemeteries in Normandy and northern Europe. Then marriage age declines dramatically in the post-war 'never-had-it-so-good' economy, reaching an all-time low in 1969. After this it just rockets up, but we'll deal with that in a moment.

What about the age gap between men and women who marry? For centuries it was around one year on average, but it started to widen around 1900 and has been remarkably steady at around two years for the last 100 years. In 2010 the median age of men marrying was 33 (31 for a single man, 46 for a divorced man, 63 for a widower) and for women it was 31 (29 for single women, 44 for divorced women, 56 for widows).

But that's just the average, so what about the more extreme cases? It's easy to smirk at stories of 24-year-old *Playboy* models marrying 81-year-old billionaires,[20] but large age differences are becoming more common, in both directions. Let's first look at marriages where the husband is older than the wife. In 1963, in 5% (one in twenty) of all marriages in Britain the husband was at least 10 years older than his wife. But in 2003 this gap had risen: in 5% of marriages the husband was 13 years older than his wife. At the other end of the scale, in 1963 the wife was more than 3 years older in 5% of marriages, while in 2003 in 5% of marriages the wife was at least 7 years older than her husband.[21] So although the average gap stays around two years, the diversity is increasing.

But, to be honest, modern marriage data are not so good at telling us about people's sex lives, because: (a) people tend to have sex a long time before marriage; and (b) many people don't get married at all. But we still want to know when people start cohabiting, whether married or not, as we have seen that it strongly influences the amount of sex going on.

It's tricky to find out about cohabitation as it is not officially registered, and so, as we found when discussing length of partnerships in Chapter 3, we have to rely on surveys rather than official statistics. Fortunately people are reasonably happy talking about cohabitation, and so we can get good data from general household surveys.[22] Let's look at the situation in the early 1980s, when Margaret Thatcher was Prime Minister, Barbara Woodhouse was training dogs on television, and Abba were singing 'Super Trouper'. At that time:

- around 60% of men and 80% of women had formed some sort of live-in partnership by the time they were 25
- over 90% of these partnerships were marriages
- the median age at first partnership was 24 for men, 22 for women
- 30% of married couples had lived together beforehand, for an average of 15 months.

By the mid-2000s things had changed radically: Tony Blair was Prime Minister, Christopher Eccleston was playing Doctor Who, and James Blunt was singing 'You're Beautiful'. But this was not all that had altered:

- only around 40% of men and 60% of women were living together by the time they were 25
- over 90% of these partnerships were cohabiting, reversing the pattern in the early 1980s
- the median age at first partnership had risen to 26 for men, 24 for women
- 80% of marriages were preceded by cohabiting, for an average of 3 years.

These are huge changes in one generation. In the twenty-five years between these two Prime Ministers, first marriages were delayed by an average of five years: roughly two years of this was due to starting any sort of partnership later, and the remaining three years due to cohabitation before marriage. So there has not been a 'flight from partnership' – just a bit of a delay in taking the plunge.

~~~~~~

1998: the first year in which the average couple
gave birth before they got married

~~~~~~

These statistics about partnership cannot tell us when sex starts, but can at least indicate when it starts in earnest. And the next 'official' marker is a birth – the gap between marriage and first birth (minus nine months) can be assumed to represent the period in which a woman is sexually active but not getting pregnant, either by chance or whatever means is available to her.

From the late 1500s to the early 1900s, before fertility started to be artificially controlled, the average gap between marriage and the first child was around fifteen months.[23] We know that some cases were considerably shorter: Queen Victoria married Albert on 10 February 1840 and gave birth to her daughter, also Victoria, nine months later, on 21 November, while her son Edward VII married Alexandra in March 1863 and she gave birth to Albert in January 1864.

The average ages at first marriage and birth between 1938 and 2012 are shown in Figure 40. After the Second World War couples started marrying younger, but the delay until they had their first child remained at around two years on average. The mean age at first birth reached an all-time low of 23½ in the late 1960s, and then started steadily increasing to over 28 in 2012, a year *less* than the average age at which people get married for the first time. The lines crossed in 1998, when the average age at which women had a baby was lower than the average age at marriage. Birth and marriage had been decoupled.

By putting this together with the data from Chapter 7 on age at first sex, we can see what happens to the gap between becoming sexually active and having a child. An average woman born in the late 1930s first had sex at 20, began her partnership at 21 (almost all being marriages) and had her first child at 23 to 24. These events were all squeezed up into

Figure 40: **Average age at first marriage and birth: women 1938–2012 in England and Wales**

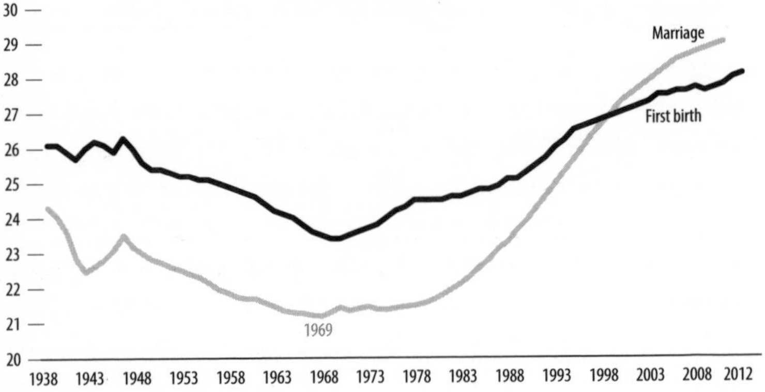

Source: Office for National Statistics.

a few years, although still longer than the 15 months in previous centuries.

But an average woman born in the early 1980s first had sex at 17, started a partnership at 23 to 24 (mostly cohabiting) and had her first birth at 27. That's ten years of sexual activity before her first birth – a massive change over two generations, and a tribute to modern contraception.

Babies and sex

3: the reduction in the expected number of babies per British woman between 1870 and 1930 (4*)

Babies, believe it or not, come from sex, and so a rather indirect way of finding out about sexual activity is to look at how many babies are born. For example, after a dip between 1914 to 1919 due to the First World War and Spanish flu, there was an extraordinary baby bubble in 1920, when 957,000 were born in England and Wales – the most in one year up to the present

Figure 41: **The Total Fertility Rate for England and Wales, 1843–2012**

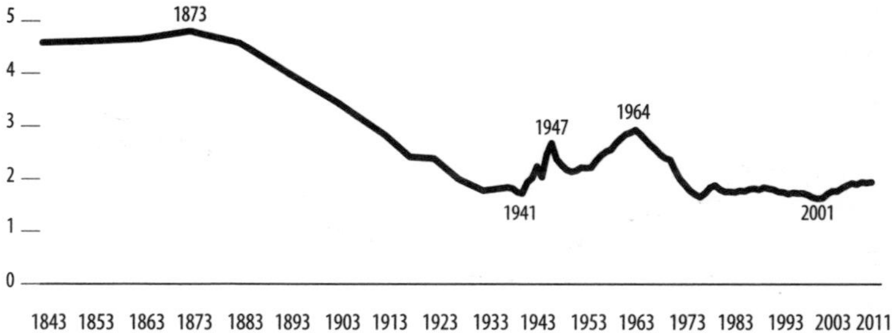

This is the number of births expected from a woman experiencing the contemporary fertility rate throughout her life. Source: Office for National Statistics are before 1938,[25] and after 1938.[26]

day, following frantic sexual activity on the return of the surviving soldiers. There was another peak after the end of the Second World War, and another in the early 1960s as behaviour became more liberal, but after the pill and legalised abortion arrived, sex and babies became less strongly associated.

Since the number of births is strongly influenced by the size of the population and by short-term events such as the end of wars, to get more insight into changing sexual behaviour we should look at the *Total Fertility Rate*, which is the number of children a woman would be expected to have, if she lived her life having babies at the rate in that year.[†]

This reveals a rather different picture, shown in Figure 41. Fertility rates go *down* fast from 1870 to the 1900s, the complete opposite trend to the total number of births. So women were rapidly reducing how many babies they had, but a lot more babies were being born! This is the kind of thing that

---

† This is similar to the idea of life-expectancy, which is the expected number of years you would expect to live, if you were subject to the current mortality rates. This is a useful but rather theoretical quantity and does not correspond to how long you, personally, can be expected to live, as future mortality rates are expected to go down.

can put people off statistics, but can be explained by a rapidly rising population.

All countries go through a fertility transition, when the number of babies changes from 'natural levels', say around 6 or 7 per family, expecting a few to die, to just over 2 and so creating a stable population – it happened in Bangladesh between 1980 and 2010, helped by contraception and increased age at marriage. It has not happened yet in Niger, where the Total Fertility Rate is still 7.

But the transition in Figure 41 prompts two questions. First, why did it happen in England after 1870? We are not particularly concerned with deep historical questions of 'why' in this book, and in any case the influences on fertility are very complex – could you say *why* you were born? Perhaps we can, possibly naïvely, say that couples realised that it was not economically sustainable to have large numbers of children in a modern industrial society. And women got fed up with always being pregnant, and wanted to give their children the best chance in life.

But second, and more relevant in our search for insights into the sexual behaviour in the past, is *how* it happened. This is one of the great historical mysteries, not quite at the level of Jack the Ripper, but still eagerly debated in scholarly texts. Essentially, did they have less sex, or did they somehow prevent sex leading to babies? We'll look at some suggestions soon, but the question brings us to one of the crucial questions of sex: if most sex is really non-procreative, how do people not have babies?

10

SEX AND NOT HAVING BABIES

As we found in Chapter 2, in Britain now, 99.9% of sex does not lead to a pregnancy. Presumably previous generations would have been rather envious of this, but, as they did not have the benefits of modern contraception, they fell back on a panoply of alternative methods to avoid their families getting too big.

Women used extended breast-feeding to space births, or blocked their cervix with various barriers: ancient methods included pessaries containing herbs such as asafoetida, and wool plugs soaked in honey or olive oil. You could also try to expel anything that did make it through: a folk belief in squatting and sneezing to avoid conception began with the Romans, and, possibly more effectively, a sponge inserted before and then douching after sex has been cited as the major reason for the decline in fertility in France in the 1700s.

Condoms played a minor role in contraception until the twentieth century, in spite of having been invented in 1564 by Gabriele Falloppio. They were originally linen sheaths held on with ribbon, and were intended as protection against syphilis ('the French disease'): the English called them 'French letters' (a term that continued into the twentieth century), and the French called them 'la capote anglaise' ('the English hood'). By the 1700s condoms were made from hand-sewn animal intestines, which could transmit sensation but were unreliable: Casonova used to inflate them to test for holes and amuse the ladies.[1] And by the 1890s there were 'rather thick and uncomfortable rubber condoms retailing at between 2 shillings and 10 shillings a dozen'.[2]

And there was, of course, always 'withdrawal', or 'coitus interruptus', which, although it does not require any equipment and is free to all, does require timing and self-control, and leaves no (statistical) trace. So let's go back to Figure 41 and the mysterious drop in fertility between 1870 and 1930. We saw in Chapter 9 that people got married later over that period, but this in itself is not enough to explain the lack of babies. People must have changed their sexual behaviour within marriage.

Contraception had been actively discussed at the start of the 1800s, after the radical views of Thomas Malthus on the dangers of population growth became established. Sponges, withdrawal and condoms became reasonably respectable topics of conversation, if only to be condemned. But historians seem agreed that devices were not widely used in Britain: condoms were expensive, and, while sponges and douching were popular with the French, they required a clean water supply. Many quack contraceptives and abortifacients were advertised, but there was limited abortion. Clelia Mosher's small sample of educated American women used a wide variety of means in the late 1800s, including douches, syringes, withdrawal and attempted timing to avoid the fertile period, but they do not appear representative of even middle-class British women.

So let's get back to the question raised at the end of the last chapter: how did Victorian women stop having so many babies? Authorities such as Hera Cook and Simon Szreter have come down firmly on the side of *continence*: they argue that there was simply less sex about, and when it happened there was often an attempt at withdrawal. This would have fitted with the advice given at the time: authorities such as William Alcott were 'quite sure that one indulgence to each lunar month is all that the best health of the parties can possibly require'.[3] The frequency of sex may have dropped in the late 1800s without people realising it – there would have been no idea what was 'normal'.[4] But it's difficult to measure what people are not doing.

The problem of measuring abstinence

15%: the proportion of women married around 1900 who used birth control, according to the Royal Commission on Population, 1949 (1*)

At last in 1914 we get some statistics, and these back up the impression of people avoiding pregnancy by simply having less sex. In the National Birth Rate Commission's fertility survey 203 out of 634 respondents said they were controlling their fertility: of these, 52% said 'continence', 13% said withdrawal, 10% said condoms (known as 'sheaths' at the time) and 18% said pessaries, douches and other 'artificial means'.[†] Although 'continence' was popular, it still was apparently only being practised by around one in six of the total sample.

But this would be a mistaken conclusion. Fortunately the data were critically analysed by a prominent statistician, Major Greenwood,[‡] who realised that the families who claimed they were not exercising birth control had just as few children as those that said they were.[5] He concluded that

---

†An unknown number of questionnaires were sent to female university graduates and their female relations who had not attended university. 787 were returned, of which 634 were useable.
‡Greenwood is still a famous person in statistics. 'Major' was his name, rather than a rank.

many families would, possibly for religious reasons, deny they were using birth control but at the same time were exercising abstinence. So the statistics point to a much larger, but unknown, proportion that should have reported that they were using 'continence'. This is a fine example of statistical detective work.[6]

This really is a treacherous area for statistics. The Royal Commission on Population† reported in 1949 that only 15% of women who married between 1900 and 1909 had used birth control, almost all withdrawal.‡[7] So, again, where were the abstainers? The historian Simon Szreter did some more statistical detective work, realising that for 'abstinence' to be recorded by the Royal Commission it had to have been for at least six months. So cutting back on sex, without eliminating it completely, would have gone unrecorded.[9] The moral of this story is that the precise wording of the question is vital: a definition hidden in the small print can make a huge difference to our view of past human behaviour. That's why statisticians are so infuriatingly pedantic.

So the data, when carefully examined, support the claim that the transition in British fertility was due, essentially, to people having less sex. And in Chapter 15 we shall use a curious set of 4* data to argue that sexual activity reached an all-time low around 1900.

---

† 3,281 women were interviewed between 1946 and 1947. The women were contacted through hospital wards and general practices: there was a bias towards urban (London and Glasgow) women who married young and were less 'well-to-do'.

‡ Szreter and Fisher's Blackburn interviewees, when referring to withdrawal, would say 'he was very careful – he would always get off at Mill Hill'. Mill Hill was the last stop on the railway into Blackburn before the main station.[8]

The importance of taking the pill

26,000: the estimated additional births and abortions in 1996 attributable to a misguided warning about the contraceptive pill

By the 1940s research into human hormones was giving rise to some extraordinary statistics: for Adolph Butenandt to get enough androsterone to cover the head of a pin, he had to start with nearly 4,000 gallons of urine, while to get one hundredth of an ounce of pure testosterone, Ernst Laqueur required nearly a ton of bulls' testicles.[10] But all this work (and all those unfortunate bulls) led to the contraceptive pill and the revolution in human sexual behaviour of the late 1960s and early 1970s: out of women born in 1946, 20% were using the pill when aged 20 in 1966, and 70% when 28 in 1974.[11]

The importance of the pill was highlighted when, on 18 October 1995, the UK Committee on Safety of Medicines sent a letter to 190,000 general practitioners informing them that new evidence suggested that 'third-generation' pills could roughly double the risk of venous thrombosis (blood clots) in the legs. No numerical information was provided, the letter took days to reach some GPs, and the media had a field day. Even though the publicity said women should continue taking their pills, one GP reported that 12% of users stopped on the day of the announcement.[12]

It is impossible directly to attribute subsequent events to women stopping effective contraception, but there is strong evidence that there was a massive impact. Conceptions in England and Wales had been steadily falling from 1993 to 1995, but rose by 26,000 in 1996. Abortions had also been falling, but rose by 13,500 from 1995 to 1996: an 8% increase. That's at least 12,500 extra births and 13,500 extra abortions owing to a single panicky letter.

And for essentially no reason. The risk of venous thromboembolism may have doubled compared with second-generation pills, but only from 1 in 7,000 women per year to 2 in 7,000, a classic example of the *relative* risk (the doubling) being apparently high but the *absolute* risk (1 in 7,000) being low: twice 'not-very-much' is still 'not-very-much'. The condition is also rarely fatal. And, if the contraception failed, the risks of thromboembolism when pregnant were 4 in 7,000. The Committee on Safety of Medicines later reversed its warnings and said the brands were fine to prescribe,[13] but suspicion of the pill's safety has lingered on.

Contraception now

21%: the proportion of 16- to 24-year-old women using emergency contraception each year (3*)

Since so much sex is non-procreative, couples must spend much of their sexually active life protecting themselves against pregnancy. Nearly 2,400 women aged between 16 and 54 interviewed for the 2010 Health Survey for England can give us an idea of how they do it.[†]

Overall, 82% of women between 16 and 54 were sexually active, and 83% of these were using some type of contraception: most of the 17% of women not using contraception were either pregnant, wanting to become pregnant, postmenopausal or possibly infertile, depending on their age. However, over 3% were judged to be at risk from unplanned

---

†The Health Survey for England 2010 sampled 8,736 addresses, and 66% of households agreed to participate: all adults in the household were interviewed simultaneously. The sexual health questions were in a self-completion booklet and were completed by around 90% of men and 92% of women aged 16–69, for a total of 2,479 men and 3,201 women.

Figure 42: **Contraception currently used by sexually active women**

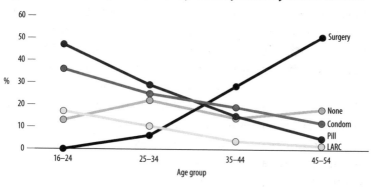

Source: women interviewed in the Health Survey for England 2010 ('LARC' = 'long-acting reversible contraception').

pregnancy, whether because they did not like contraception or found it unsatisfactory, their partner did not want to use contraception, or other reasons.

The type of contraception used varies hugely depending on the age of the woman, and Figure 42 shows the results for sexually active women in four different age groups.[14] The pill is used by 22% of sexually active women overall, but by nearly half the younger respondents. Similarly, male condoms are used by 22% of sexually active couples overall, again with a sharply declining age profile. In contrast, surgical methods (sterilisation of either partner) is used by over half the oldest group but by very few of the youngest.

Long-acting reversible contraception (LARC) – whether implants, injections or patches – is used by 7% overall and by nearly 20% of 16- to 24-year-olds. LARC is a public health priority in the USA, particularly for younger and poorer women – when offered free to 10,000 women in St Louis, 67% chose this method.[15] LARC is free under the NHS, but in the USA it will cost more than $1,000 if you are uninsured, and so is beyond the reach of many who would most benefit.[16]

Other methods are used more rarely: coils/intra-uterine device (IUD) by 5%, withdrawal by 2%, 'natural' methods (timing of sex to avoid fertile periods) by 3% and our old friend abstinence by 0.6%. The emergency contraceptive ('morning-after') pill had been used in the preceding year by 7% overall, and 21% in the 16–24 age group – that's more than 1 in 5 young people seeking emergency contraception each year. But only 1% had used it twice in the preceding year, which suggests they were not making a habit of it. Again there are payment issues in the USA with this method, with the Supreme Court ruling in 2014 that companies that provide health plans cannot be compelled to pay for emergency contraception, as some consider it abortion. Sexual health easily becomes a political issue.

The effectiveness of contraception

55: the number of times the condom broke or came off in 3,715 uses under 'perfect' conditions

We shall see in the next chapter that nearly half of all pregnancies in the UK are unplanned or ambivalent, and so presumably came as a surprise, and not always a welcome one. And yet over 80% of sexually active women are using contraception, and so many, if not most, of these pregnancies occur when some sort of birth control has been used. So how reliable are these different contraceptive methods?

If you think you are what is known as a 'perfect user' of contraception, the NHS Choices website will tell you the success rates of different types of contraception, in terms of the estimated proportion of women who will not be pregnant at the end of a year's use, and these look very encouraging.[17] But not everyone actually is perfect, and so Figure 43 also shows a more realistic assessment of what the ordinary fallible human can expect.

There are some important features of this graph. First, it

184

Figure 43: **Contraceptive failure rates under 'typical' and perfect use**

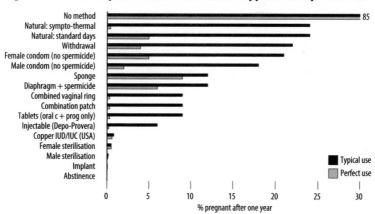

Source: Trussell 2011.

reports *failure* rates rather than *success* rates – and a 5% failure rate tends to sound worse than a 95% success rate. Second, it includes failure rates for 'typical use', which is supposed to allow for those all-to-human characteristics of forgetting, drunkenly fumbling and so on.

We immediately see that the techniques can be divided into two broad categories: those that don't require human skill or judgement, which have low failure rates in theory and in practice – fortunately a man does not have to use a vasectomy according to the instructions. And those where the capacity for human error or equipment failure is all too obvious, such as using condoms and withdrawal.

These are the base figures reported everywhere from NHS Choices to Wikipedia, but where do they come from? What sort of 'perfect' and 'typical' people allow their contraceptive failures to be studied? They are based on the work of James Trussell, Professor at Princeton University,[18] and close scrutiny reveals a judicious combination of data and considerable judgement – the term 'guess' features repeatedly in his analysis.

Let's first look at the most commonly used methods in

the UK: male condoms and the combined oral contraceptive pills. The annual failure rate of male condoms is claimed to be 2% in perfect use, but 18% in practice. The 'perfect use' figure comes from experiments in which volunteer couples were randomly allocated to two groups: one to use standard latex condoms and the other experimental versions. Three of these studies were put together to estimate failure rates, providing an interesting insight into condom use under test conditions.[19]

Out of 3,715 attempted uses of standard condoms, around 1% (38) did not even make it to the starting line; 16 broke when unwrapping, 10 could not be put on and 3 were put on the wrong way round. But even once they'd been correctly applied, there were 55 failures (1.5%) when being used: 16 broke, 16 slipped off during sex and 22 slipped off on withdrawal. In 2,248 monthly cycles when the condom was used for all sex, there were 4 pregnancies: that's about one-sixth of 1%. Twelve times this monthly rate gives the 2% for a year shown in Figure 43.

These 'perfect' users may not seem particularly skilled, even clumsy, but 'typical' use is considerably less controlled: it does not even mean that condoms were always used, just that the women in the survey considered that they were using them. Trussell uses data from the US National Survey of Family Growth (NSFG) surveys of contraceptive practice – in the 2002 survey of 7,643 randomly chosen women, 14% (1 in 7) of condom-users said they were pregnant after a year: this was raised to 18% after allowance was made for women not reporting abortions.[20] Factors which tended to increase the failure rate included being under 30, being black, intending to have more children, cohabiting, being poor or already having children. So users without those characteristics would have a far lower failure risk than 18% – maybe even down to the 'perfect' 2%.

For the combined oral contraceptive Trussell says that

there is some evidence of occasional failure even if perfectly used, and so he judges the annual failure rate of 0.3%, or once in 300 years. In typical use, allowing for mix-ups on dates, poor memory, throwing up and so on the same surveys suggested 9% would be pregnant after a year – that's one pregnancy for every 11 years' use. At these rates, given the 3 million women using the pill in the UK, that would mean around 270,000 unplanned pregnancies each year through failure of the pill, which seems excessive. But again, these risks are only average: they are higher for women under 30, intending to have more children, unmarried or already with children, so the 9% figure may be too low for those with these 'risk factors' and far too high for many people without these characteristics.

We can compare these to the old medieval stand-by – withdrawal or coitus interruptus. Here the expert admits that his estimated 4% annual failure rate under 'perfect' use of withdrawal, presumably by a master of timing and restraint, is a 'guess'. He assumes that no sperm comes from an ejaculation, but that any risk is attached to sperm in pre-ejaculate: some studies had found no evidence of this, but one that examined samples within two minutes of 'production' found at least some motile sperm in 37% of samples.[21] This study collected 40 samples of pre-ejaculatory fluid from 27 volunteers: 10 of the subjects produced fluid with a small, but measurable, amount of motile spermatozoa, but none of the others did, leading to the conclusion that withdrawal might be a perfectly fine method of contraception for some couples, but not for others. So it depends on the man, not so much the occasion.

The failure rate from 'typical use' of withdrawal was estimated to be 22% per year from an average of the 1995 and 2002 NSFG surveys: this is increased by the usual factors – being young, poor and so on – but particularly for women who have never been married, possibly owing to the difficulty of negotiating such a technique with a partner with whom the prospect of having children may not have been discussed.

But how good are all these data? Although based originally on randomised trials and good surveys, we've seen that many assumptions and judgements are required to get to the numbers in Figure 43. I would only rank them as 2* statistics.

More importantly, do these numbers apply to *you*? For example, typical use of withdrawal is credited with a 22% annual failure rate, and so over ten years you might estimate that you had a 92% chance of getting pregnant.[†] Not great if you are trying to avoid this event.

But suppose, to take an extreme and perhaps implausible example, that the '22% annual failure rate' means that 22% of men are hopeless at withdrawal, but 78% of men are meticulously careful and have no sperm in their pre-ejaculate. That means that after ten years 22% of partners will have got pregnant (at least once), but the remaining 78% will still be protected. That's 22% failure over ten years, rather than 92%. This is a somewhat subtle statistical point that arises when the 'risk' varies widely across people, and so the 'average' risk does not represent anyone's actual experience.[22]

So whether or not your contraception works depends in a complex way on your fertility, competence, age, circumstances and so on – and the published statistics can never be more than a guideline.

[†]The chance of being pregnant after 10 years = 1 – the chance of not being pregnant = 1 – (chance of not getting pregnant each year)^{10} = $1 - (1 - 0.22)^{10} = 0.92$.

Abortion

90,000: estimated number of illegal abortions
in Britain each year in the 1930s

So, if your contraception does fail, what do you do next? The history of termination of pregnancies stretches back thousands of years, whether induced by herbs, drugs or mechanical means.[†] Statistics are sparse and unreliable: we know that abortifacient drugs such as diachylon were used extensively in the late 1800s and early 1900s for 'bringing on periods'. It's been argued that the highest abortion rates occurred in the 1930s, with widely varying estimates of 60,000 to 125,000 a year[23] – the accepted figure is around 90,000 a year at a time when there were 600,000 live births annually.[24] This then illegal practice may have accounted for around 13% of pregnancies: an extraordinarily high figure.

After a long and heated argument abortion became legal in the UK in April 1968. By 1973 there were 167,000 each year, although a third of these were non-residents who came for their abortions, many from Ireland. In 2013 there were 185,000 registered abortions in England and Wales, which was around one in five of all pregnancies[25] – about double the rate estimated in the 1930s, when it was illegal. Nearly all were under three months' gestation, and around half were medical abortions induced by drugs (mifepristone): in 1991 only 4% were medical.

Figure 44 shows the distribution of numbers of abortions to residents of England and Wales by the age of the woman.

[†] A last resort has been abandonment of the new-born: in ancient Greece and Rome the father had the choice of exposing a child, left like Oedipus in a place specified for the purpose. The Coram Foundling Hospital in London carried on that role and cared for 25,000 abandoned children between 1741 and 1954.

Figure 44: **The number of abortions to residents of England and Wales, 2013**

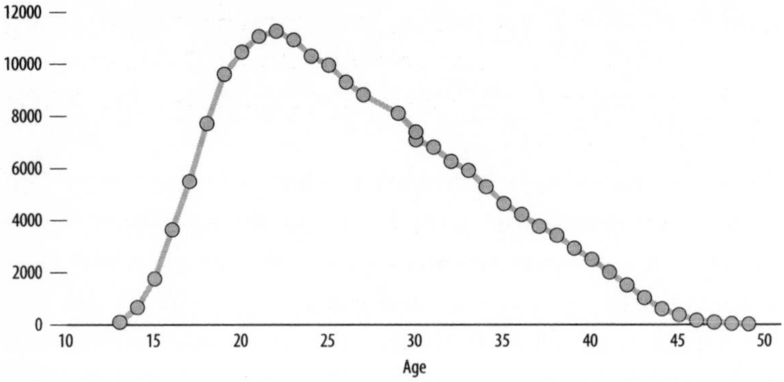

Source: Department of Health.

The peak is for 22-year-olds: overall, 1.6% of women between 15 and 44 had an abortion in 2013, the lowest rate for sixteen years. As we saw in Chapter 7, more than half of all pregnancies in adolescents aged 15 end in abortion, and this figure drops steadily until, for women aged 30–34, only around one in ten pregnancies ends in abortion. Then the rate rises again – around one in three pregnancies to women over 40 ends in abortion.

The majority (81%) of women having abortions are unmarried, but of these around two-thirds have a partner, although we don't know the extent to which a joint decision has been made. And for over a third of cases (37%) this was not their first abortion.

Yet again we are faced with the utter inadequacy of these crude statistics to convey the complexity of these 185,000 individual stories.

The ways in which women have sought to control their fertility have been, until recently, rather a mystery, and the paltry statistics that have been collected have tended to be unreliable and based on asking the wrong questions. Even now the estimates of the effectiveness of contraception seem fairly ropey. This is an area where averages seem particularly weak guides; it's hardly surprising that your chances of getting pregnant depend crucially on your individual behaviour.

The statistics can show us the big picture, but the details are left for us to fill in.

11

SEX AND HAVING BABIES

~~~~~~
34%: the claimed annual fertility rate of 35- to 39-year-old women
not using contraception, based on 300-year-old data (1*)
~~~~~~

Jean Twenge left it until she was 35 to try to have a baby, only to be told that only a third of women in their late 30s will get pregnant in a year of trying, and 30% will never manage it at all. She was shocked. Following many women who have valued their career, she asked herself: have I left it too late to have a family?

If you consult the 2013 guidelines on fertility from the UK National Institute for Health and Care Excellence (NICE),[1] you will see the kind of numbers that Jean was told: NICE's graph, reproduced as Figure 45, shows that the annual fertility rate of partnered women aged 35–9 who are not using contraception is 34%.

Convincing? Well not if you dig deeper, as Jean fortunately did.[2] The crucial insight is that these worrying figures are based on historical data – in fact, very historical indeed. We've already used family reconstitution studies to look at pre-nuptial pregnancy and other behaviours from centuries ago: the originator of this method was the French historian Louis Henry (1911–91), who had the grand ambition of

Figure 45: **Claimed annual fertility rate, from NICE guidelines**

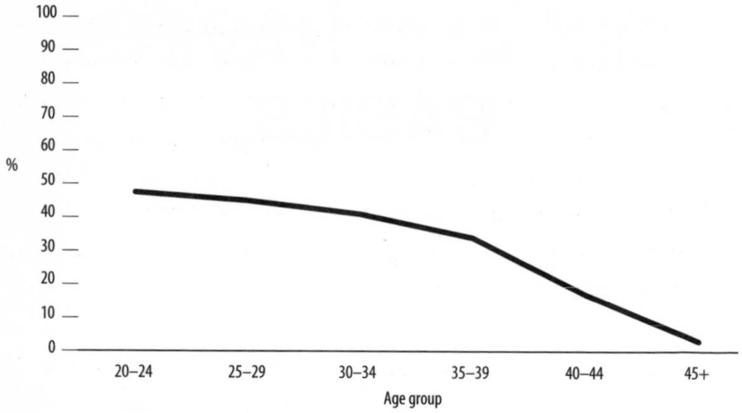

Based on Heffner (2004)[3]

reconstituting the history of all the individuals in France. As a start he put together histories from rural parishes between 1670 and the Revolution in 1789, and civil registrations from 1789 to 1830, the year of the next, but smaller, revolution.

Henry's data were used to estimate the 'natural fertility' rate of women living without access to contraception: for example, 300 years ago, around 48 out of 100 rural French women aged between 20 and 24 would give birth every year, compared to 17 out of every 100 40- to 44-year olds. All very interesting for the historian (and the statistician). But, rather amazingly, these fertility figures, put together with additional registers from Geneva in 1600 to 1649, Canadian marriages 1700 to 1730 and so on are still being quoted by NICE and other much-cited sources.[4]

So would these figures have been relevant to Jean Twenge in 2007? For a start, many of these women from the 1700s would still have been breast-feeding from a previous birth, and this would have eliminated or reduced their fertility. Then older women might well have already had six or seven children, and might have become sterile owing to complications of childbirth. They would have led hard lives with minimal healthcare. But crucially, if women did not want more children, maybe they were just avoiding sex.

So even if the old French data do represent the fertility rate of married women who were not using modern methods of contraception, they may have limited relevance, if any, to the fertility of modern women, who have delayed having children until an older age but then want to get pregnant. Such as Jean, who went on to have three children after 35.[5]

But to get pregnant, it helps to have sex. So how much do you need to have?

The chance of getting pregnant from having sex once

53%: estimated peak chance of getting pregnant from a single 'coital act', for an average woman aged between 19 and 26 (3*)

First, a bit of jargon. The chance of getting pregnant in one month, assuming no contraception is used, is known as 'fecundability': it is usually taken to be around 15% to 30% in different populations. But this is the average: for particular couples it can vary far more, depending on who they are and, of course, what they do with each other. And, sadly for those who very much want a baby, for some it will be 0%.

Estimating fecundability is difficult. We've just seen how data from hundreds of years ago can be dusted down so as to get a description of 'natural fertility'. But when giving advice now, we should be interested in women who are really trying to have children. The irony is that this can indeed be roughly estimated from historical data using more sophisticated analysis, where the trick is to look at births soon after marriage, when we might assume there is some serious sexual activity and a wish to have children. The Cambridge group used English parish data from 1580 to 1837 to estimate the monthly chance of pregnancy resulting in a live birth for women between 35 and 39 to be 10%, which corresponds to a 72% chance of a birth over a year.[6] Much higher than the claimed fertility of 34%.

The alternative, rather more appropriate, way of estimating fecundability is to study modern couples who are deliberately choosing not to use artificial contraception. The 'European Study of Daily Fecundability' was an extraordinary and influential exercise that recruited 782 couples that were attending seven Natural Family Planning Centres – most did not wish to get pregnant, although apparently their intentions often changed as the study progressed. Each woman was asked to record her daily temperatures and when sex took place.

Some 6,724 cycles were monitored, an average of 8.5 for each couple, including 3,175 with intercourse. There were a total of 487 pregnancies – that's 62% of the couples – so the 'natural planning' was not very successful if they really didn't want to get pregnant. The day of ovulation was assessed by the 'three over six' rule – when three daily temperature measurements were all higher than in the six preceding days, the last day of the lower measurements is assumed to be the day of ovulation. The couples recorded an average of 6.2 'acts' in cycles that led to a conception, and 4.5 in the other cycles[7] – roughly in line with the median frequencies for younger people shown in Chapter 2.

It's long been recognised that the time of peak fertility is around ovulation: the rhythm method of contraception involves avoiding this time. But how does the chance vary across the whole month? This study allowed these chances to be estimated by labelling each 'act' by its day relative to

---

† If there is a 10% chance of getting a successful pregnancy each month, then the chance of not getting pregnant in any month is 0.9. The chance of not getting pregnant over twelve months is, assuming they are independent, $0.9^{12} = 0.28$ or 28%. Hence the chance of getting pregnant and giving birth over the whole year is 100% – 28% = 72%.

Figure 46: **Average % chance of conception from a single act of intercourse**

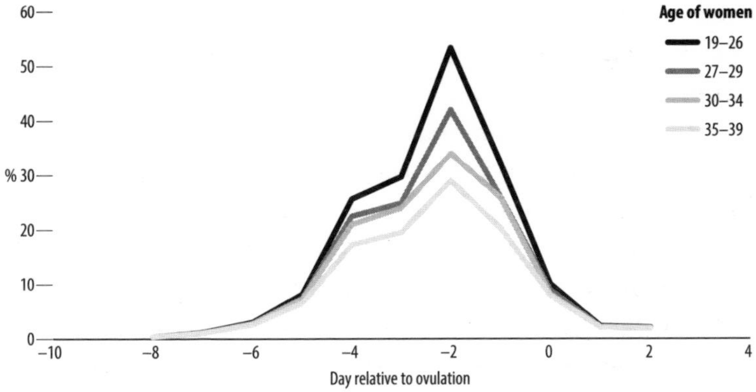

Source: European Study of Daily Fecundability.

ovulation: −2, −1, 0, 1, 2 and so on, where −2 would correspond to two days before ovulation. If there is only one act in a cycle, followed by pregnancy, then it's easy to see when the crucial event actually happened. But it's unusual to observe such cycles: previous studies have tried to persuade couples to have sex only once a month but had a high drop-out rate, which perhaps isn't much of a surprise.

So a fancy statistical model was used to estimate the probability of conception on any day, taking into account the age of both the woman and the man, and also allowing for variability between couples as to their fertility.[8] Figure 46 shows the results. Women aged 19–26 had, on average, a peak probability of conception of around 53% two days before ovulation, but this peak drops to around 29% for women aged 35–9.

The probability of conception was only above 5% for the six days ending with ovulation day, and this overall interval of fertility did not depend on age. Over the whole twenty-nine days of the cycle the average chance of conceiving on a particular day was around 6% for 19- to 26-year-old women, and 4% for a 35- to 39-year-old. So, if you can imagine a single random act of sexual intercourse (and such things can happen, so I've heard), that's the average chance of getting pregnant. But if an average, but extremely enthusiastic, young couple have unprotected sex every day of the cycle, then there is an 88% chance they will conceive, and it is more likely than not that this will have already happened before the day of peak fecundability.[†]

Of course, all these figures are averages, and unsurprisingly the man also plays a role. Men above 35 reduce the odds considerably: for a woman of 35, the peak probability was estimated to be 29% for a partner the same age, but this goes down to 18% if the man is aged 40.

The researchers also found a lot of variability between couples for reasons they could not explain: for example, they estimated that, for 27- to 29-year-old women, 1 in 20 would have a peak probability of conception below 5%, and, at the other extreme, 1 in 20 would be extraordinarily fertile, with a peak probability of conception of 83%. So those stories about couples who appear to just need to have sex once to have a baby are not necessarily exaggerated.

It's clear that older couples have a lower chance of conceiving, but is this because they are becoming completely sterile, or just less fertile as they get older? And if the latter, how much harder will they have to try? The data from the 782 Natural Planning couples were re-analysed to try to answer these questions, using a 'mixture model' that allows for a proportion of couples who will never have a baby no matter what they do.[9] This proportion that were effectively sterile was estimated to be around 1%, although this will be on the low side, as the study only goes up to age 40 and did

---

[†] In fact, the chance that they will have conceived before the peak day is 54%. This means that, for people having a lot of sex, the distribution of the day of conception is shifted to the left of the pattern shown in Figure 46 – this becomes important in Chapter 15, when we come to the habits of servicemen home on leave at the end of the Second World War.

not recruit people with known fertility problems. The important finding was that this 1% was independent of the age of the couple, which suggested that the decreases in fecundability with age were due not to couples becoming completely sterile, but just having reduced fertility.

The detailed estimates of daily chances of conception mean that we can predict the time to pregnancy according to different behaviours. With sex twice a week, it's estimated that 92% of 19- to 26-year-old couples will conceive within a year, and 98% within two years. This compares to 82% within one year and 90% within two years for 35- to 39-year-olds.

These chances drop somewhat if the couples have sex only once a week: for example, the 19- to 26-year-olds only have an 82% chance. However, there is little advantage of going from twice a week to three times a week, so they are allowed a break. The effect of a man being 40 rather than 35 is about the same as dropping sex rates from twice to once a week.

So some couples will have to wait longer than others, but there is great variability between couples whatever their age. The researchers conclude that couples should not give up after a year of trying, even though this is a standard measure of 'infertility': they estimate that of couples comprising a woman aged 35–40 and a man aged 40, who have not conceived after a year of trying, 43% will conceive after two years.

All this is in marked contrast to Figure 46. But, to be fair, NICE also provide the modern estimate that 82% of 35- to 39-year-olds will expect to be pregnant within a year,[10] and it's a mystery why they continue to publicise the fertility of old French peasants as if it has some relevance today. And the major lesson from all this analysis is that behaviour is important: after years of careful research I think we can confidently conclude that, if you want a baby, then it's a good idea to have sex.

How many babies can a man sire?

600: the number of sons of Moulay Ismaïl Ibn Sharif,
Sultan of Morocco, as reported in 1704 (1*)

Known as 'the bloodthirsty', owing to his habit of mass beheadings, Sultan Moulay Ismaïl Ibn Sharif of Morocco ruled from 1672 to 1727 and is reputed to have had the most number of children of any man in history. A French diplomat, Dominique Busnot, reported that by 1704 he had 600 sons by 4 wives and 500 concubines, which works out at around 1,200 children over his 32 years in power (although the daughters of the concubines were suffocated at birth). I would only consider this a 1* number at very best, but is it even feasible for someone to have so many children, even with access to a harem that size?

A recent study tested the plausibility of this claim using computer simulations under a range of different behaviours.[11] Under two assumptions about fertility across the cycle, the authors estimate that the sultan would have had to have had two 'copulations' (to use the researchers' term) a day for 32 years – that's around 23,000 times – in order to have 1,200 births over that period. This works out at an overall rate of 'child per copulation' of 5%. Another set of assumptions suggested that he only needed to have had sex once a day, corresponding to a 10% success rate.

The maximum age of the concubines was 30, and we previously estimated an average probability of a conception of 6% from a 'random copulation' with women that age. I therefore believe that the sultan would have needed to have had sex twice a day for 32 years to have had the reputed number of children: if you don't think this is plausible, then you should conclude that Dominique Busnot was being spun a tall story in 1704. The computer simulations could also have told the sultan he did not need to go the expense of having 500 concubines, as he could have obtained the same fertility from around 100. But then he would not have had such a wide choice.

Unplanned pregnancies

45%: the proportion of pregnancies that Natsal-3 estimated were either 'unplanned' or 'ambivalent' (3*)

So, you've had sex, and you or your partner is pregnant. How do you feel about it? Was it a deliberate, carefully planned event?

Statistics about 'unintended' pregnancies abound: figures around 50% are often quoted for both the UK and the USA. But what does 'unintended' mean? It's not a black or white issue – there may be a degree of ambivalence, such as thinking that a baby would be nice to have some time, but perhaps not right now. For example, for births (rather than pregnancies) in the USA, the National Survey of Family Growth estimated that 23% of births were actively 'unwanted', and 14% were 'just' mistimed.[12]

The Natsal team have tried to deal with this by developing a more subtle scale to measure the degree of unplannedness – the London Measure of Unplanned Pregnancy, shown in Figure 47.[13] The points are added up, and you can score between 0 and 12. The scale could be chopped up in all sorts of ways, but Natsal classify the pregnancy as 'unplanned' with 3 or fewer points, 'ambivalent' with between 4 and 9 points, and 'planned' for scores of 10 or more. So if you score five 2s, you are allowed one 0 and it still counts as 'planned'.

In the Natsal-3 survey, women were asked about any pregnancy in the last year, and data were obtained on 590 women.[14] The median score was 10, suggesting the 'average'

Figure 47: **The London Measure of Unplanned Pregnancy**

Q1 At the time of conception
0.   Always used contraception
1.   Inconsistent use
2.   Not using contraception

Q2 In terms of becoming a mother
0.   Wrong time
1.   OK but not quite right
2:   Right time

Q3 Just before conception
0.   Did not intend to become pregnant
1.   Changing intentions
2.   Intended to become pregnant

Q4 Just before conception
0.   Did not want a baby
1.   Mixed feelings about having a baby
2.   Wanted a baby

Q5 Before conception
0.   Had never discussed children
1.   Discussed but no firm agreement
2.   Agreed pregnancy with partner

Q6 Health preparations* before conception
0.   No actions
1.   1 action
2.   2 or more actions

*Health preparations include: taking folic acid supplements, stopping or reducing smoking, stopping or reducing alcohol consumption, healthy eating and seeking medical advice before. Source: Natsal-3.

birth was just about planned, and 50% of pregnancies scored between 5 and 12.

Overall, around 16% of pregnancies were classified as 'unplanned', 29% 'ambivalent' and 55% 'planned'. But Figure 48 shows this strongly depends on age: for 16- to 19-year-olds only 12% were 'planned', compared with 68% for 30- to 34-year-olds. This still means that a third of pregnancies to older mothers were unplanned or ambivalent, though, so this is not just an issue for the young. The rates of unplanned pregnancies also increase for smokers, those with lower education, first sex at an earlier age and so on.

Interestingly, only 67% of the recorded abortions arose from unplanned pregnancies: 23% of abortions were after ambivalent pregnancies and 10% from planned pregnancies. This shows one cannot simply equate abortion with unintended pregnancies, although these are only rough

Figure 48: **For different age groups, the percentage of pregnancies that were unplanned, ambivalent and planned**

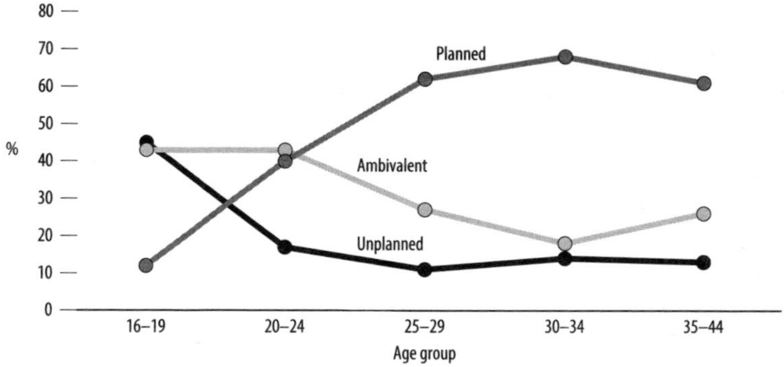

Source: Natsal-3.

estimates, as only 80 abortions were recorded. Some events stand out: of pregnancies of truly single women, with no partner or cohabitee, one in six was planned. There must be many powerful stories behind these bald statistics.

These data rely on self-reporting about events up to a year ago, and, of course, feelings can change: what might feel like a shock at the time can, in hindsight, turn into an unexpected gift. Babies born after 'ambivalent' or 'unplanned' pregnancies may be as cherished as those that were the result of meticulous organisation. But if we apply the Natsal estimates to Britain in 2011, it would mean 540,000 planned pregnancies, balanced by 285,000 ambivalent and 160,000 unplanned.

That's a lot of surprises.

9%: the excess in number of births in September compared with April (4*)

So let's pretend you're pregnant, you're pleased this has happened, and you're waiting for your baby. I bet it's going to be a Virgo.

Why should I guess that your little bundle is going to

Figure 49: **Relative numbers of monthly births and conceptions in England and Wales**

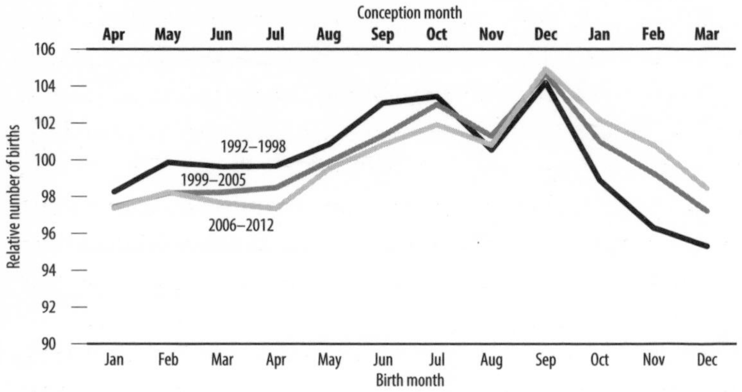

Baseline = 100. Source: Office for National Statistics.[15]

arrive in September? Well, it's a bit of a long shot, as it has to be based on understanding when couples have sex, and therefore stand a higher chance of conceiving. Do they just do it when there's nothing else to do? Or maybe it is influenced by the time of year: people may get together as the sap rises in the spring, or while keeping warm on those long winter nights.

As we can't directly see what people are doing, perhaps we can infer it from what it gives rise to. So if we look at when babies are born and subtract nine months, we can work out roughly when the crucial act was performed. So have a look at Figure 49, which shows the relative number of births in each month in England and Wales between 1997 and 2012.[†]

This shows a clear and consistent pattern. The low-spot for births is from January to April, corresponding to conceptions

---

† We can't just compare the total number of births each month, as February will inevitably look low as it has three days (10%) fewer than, say, December, and we will search in vain for reasons why May (i.e., February minus nine months) is such a sexual desert. So the *daily* rate of births is calculated, and then each month is assumed to have 365.25/12 days = 30.4 days.

between March and July – so much for the sap rising in the spring. Then from April there is a steady rise, apart from a small dip in August, until the obvious peak month for birth is September, suggesting that for some reason December is a particularly fertile month. In fact, according to a story in the *Daily Mail* headlined "'Tis the season to conceive!', Jo Hemmings, relationship expert for We-Vibe ('The world's no. 1 couples vibrator'), says that if you want to conceive, you should 'Get out those stockings and suspenders, which probably haven't seen the light of day during the summer months.'[16] Let's hope you've got central heating.

After the apparent sexual exuberance of the Christmas holidays, there is a steady decline back to the spring. In 2012 this behaviour amounted to 63,000 births in September compared with 58,000 in April – nearly 9% higher. Of course, some particularly organised couples could have been conspiring to have their child born in the autumn so they would be top of their class: as we shall see later, being older brings clear advantages at school. But is this pattern common to other countries?

Using data from the USA, we can look at births for individual days between 1969 and 1988, shown in Figure 50. There is a similar peak in September and a low period from January to April. These daily data allow us to see spectacular drops in holidays – with fewer born around 4 July and very few on Christmas Day. The burst in births just before the New Year has been blamed on money-minded parents seeking to have births induced in order to get tax breaks that would be lost if the baby were born in January. But it is probably just making up the backlog that has built up through delaying births over Christmas.

This autumn peak in births is a modern phenomenon. We have reasonably good data back to 1540 from the registers of 404 English parishes:[17] Figure 51 shows when baptisms occurred over the next three centuries. This is a dramatically

Figure 50: **70 million birthdays in the US between 1969 and 1988**

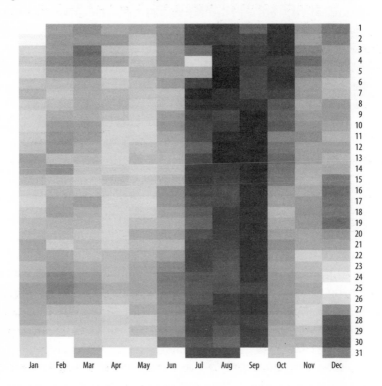

The 128 that were reported to have been born on impossible dates such as 31 April, 31 September and so on have been removed. The darker shading indicates more births

different pattern from the present day. Baptisms were higher than average in the first four months of the year and then plunged down in summer. And compare the vertical axis with that of Figure 49: the historical variation is far larger than now, with nearly 50% more baptisms in March than July. As there was no artificial contraception, this must have been the consequence of some extraordinary changes in sexual behaviour over the year, with May and June apparently a time of either energetic activity or a sudden throwing of caution to the winds.

This striking pattern was common over northern Europe at that time. The effect meant there were fewer births during

Figure 51: **Relative number of monthly baptisms in English parishes during different periods, and corresponding conception month**

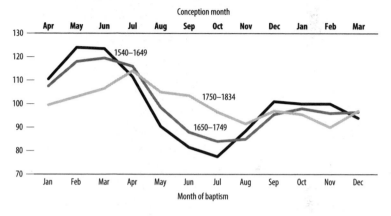

Baseline = 100.
Source: Cambridge Group for the History of Population and Social Structure.[15]

summer crop-gathering, which presumably improved productivity, but the increased birth rate in the cold months must have contributed to the high infant mortality.[†] The drop in conceptions between August and November is very steep – apparently there was little sex in the autumn, which could be to prevent births in summer when there was hard work to be done in the fields, although the same pattern is shown in London.

Maybe this pattern – for first births anyway – is because of when marriages took place? The register shows there were strong peaks in marriages around May and June, which were slacker work periods, and so subsequent births explain some of the seasonal cycle. Marriages also peaked around October, which included autumn 'hiring fairs', where workers had to choose whether to commit to a contract or plunge into marriage. Marriage showed a late summer trough in August, at

[†] This study only has baptism records, rather than birth dates, but in the 1500s baptism followed soon after birth. Later some baptisms got more delayed, which contributed to the flattening out of the pattern.

peak harvesting, and very low rates in March and December, reflecting pre-Reformation prohibitions on marriage in Lent and Advent that continued for centuries.[†]

But it was not just June when sex was in the air: there is also a suggestion of increased sexual activity in December, which may be early evidence of the 'Christmas holiday effect' suggested above. And this does seem the most plausible explanation for the current pattern of a September peak in births, backed up by data on condom sales, rise in sexually transmitted infections and abortions.[19] In the southern hemisphere, where the seasons are reversed, Australia provides a useful comparison. After adjusting for the length of the month, September is the clear peak in births, again pointing to the influence of the Christmas holidays. But there is a secondary peak in March, suggesting a separate effect of winter nights in June.

In Britain the combined effects of winter nights, holidays, alcohol, Christmas bonhomie and bad television clearly create a heady sexual concoction. But what about the repeated claims in the media about major black-outs leading to a spike in births nine months later, presumably because, if you can't watch *The Killing 3* or send emails, you are forced to acknowledge each other's presence? Given the explanations we've seen for the drop in sexual activity over the past decades, perhaps this is not implausible.

So after Hurricane Sandy in late October 2012 there was the inevitable headline in July 2013 'Sandy baby boom begins! New Jersey hospitals report that the number of deliveries are up by 30% nine months after the Hurricane.'[20] Similar stories

---

[†] It's been estimated that in England in the late 1500s the restrictions to marriage covered around a third of the year. It all got a bit absurd: as Easter is a movable feast, in about two years out of every ten it was only possible to marry, over the four months between December to March, in one week in January.[18] It is hardly surprising the Advent prohibitions started breaking down by the 1600s.

come after every hurricane, and just as quickly are debunked by demographers as urban legends.[21]

Disasters do not encourage sex, but simple lack of electricity might. During power outages in 2009, Uganda's Planning Minister, Ephraim Kamuntu, was reported to say 'While the rest of the world is working in shifts, we in Uganda are going to bed early. Then we complain that the population is growing. Why not?'[22] In fact, Africa has provided some reasonable evidence of baby booms. In 2008 an electricity supply cable in Zanzibar broke, so that some villages had a supply and others didn't, creating a natural experiment that proved irresistible to researchers. The subsequent month-long black-out was associated with a 20% increase in the number of births eight to ten months later in the blacked-out villages. When the households were surveyed, they found the black-out had altered leisure behaviour, with more time spent in the home, presumably quite a lot of it in bed.[23] These prolonged outages are not just a feature of Africa: there was power rationing in Colombia throughout much of 1992 after poor rainfall reduced the hydroelectric output, and careful analysis suggests an additional 8,500 babies were born over the expected 210,000 – around 4% extra.[24]

If there is an effect associated with the whole Christmas holiday, what if that celebratory feeling was packed into a single euphoric night? Such as when your favourite candidate becomes President of the United States? The night of Obama's victory in November 2008 was supposed to have led to a bubble of babies in August 2009, but once again this was dismissed by spoilsport experts. As an explanation for the lack of impact, it was pointed out that, although Obama won the electoral college by a large margin, the popular vote was much closer: although the 52% of the population that voted for him might be feeling exuberant, the remaining 48% would have been gloomy enough to cancel out any additional activity.[25]

In Europe we may not get so worked up about politics, but football can rouse serious emotions. On 6 May 2009 Andrés Iniesta scored a last-minute winning goal in injury time against Chelsea which put Barcelona (Barça) in the UEFA Champions League Final, and an informal survey nine months later led to newspaper headlines about a 45% increase in births – the so-called 'Iniesta generation'. We know that Barça have fervent supporters, but could Iniesta really have fathered, by proxy, so many offspring?

Analysis by Catalan statisticians of two counties in central Catalonia that were traditional supporters of Barça revealed that in February 2010 there were 183 births, compared with an average of 157 in 2007, 2008, 2009 and 2011.[26] This rather small excess of 26 extra births was confirmed with more sophisticated analysis, and it was estimated that there were 14% extra births in February, and 12% extra in March. But this pattern was not reflected in Barcelona as a whole, so it looks like Iniesta did not generate as many children as claimed: and, of course, none of those 183 will know whether they are one of the minority who only exist in this world because of that winning goal.

Does it make a difference when you are conceived?

28%: the increase in the chance of getting into Oxford or Cambridge for those born in October compared with those born in August

My birthday is 16 August, making me a November conception, and under the astrological star sign of Leo. I should therefore, if we give these things any credibility, be confident, ambitious, generous, loyal and encouraging: the insight of this astrology business is amazing – how could they know I was exactly like that? But reading on for a bit, I find I am also supposed to be pretentious, domineering, melodramatic, stubborn and vain: so maybe it's just a lot of superstitious nonsense.

If, like me, you think that astrology does not have great predictive ability, it may come as a shock as to how much of your future success or failure may be attributable to when you are born, and therefore depends on when you were conceived. This is well known when it comes to sporting success, since the oldest in each year-group have a distinct physical advantage. But if you want your child to do well in school and college, when should you aim to conceive? Such a simple, free choice (although not guaranteed) might outweigh the effects of an expensive education.

I was fortunate enough to be born into a supportive family with excellent free local schooling. I've always thought of myself as ludicrously privileged, but it turns out I have had a distinct disadvantage in my educational prospects. Consider my birthday, 16 August, and remember that the English school year begins on 1 September, so I was lumped together with children who were born between 1 September 1952 and 31 August 1953. I had at least two distinguishing features in school: I had the longest and most ridiculous name, and was always the youngest in the class.

Researchers have known for nearly fifty years that the younger kids in each class tend not to do so well academically, and this pattern continues – it's known as the 'relative age effect' (RAE). This can be huge at younger ages. Consider the proportion of children who reach the expected achievement level at Key Stage 1 at age 7 – for children born in August this is around 42%, while for those born in September it is nearly 70%.[27] This is a massive effect. And at the other end of the spectrum, 15% of August-born children have been identified as having Special Educational Needs (SEN) by age 11, compared with less than 10% of September-born children.

By age 16 these differences have flattened out a bit, but are

Figure 52: **The percentage of people born in each month in 1993–4 who were later admitted to Oxford or Cambridge as undergraduates in 2012**

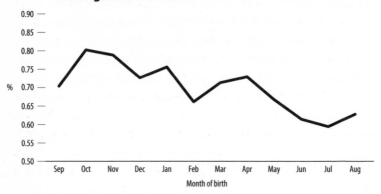

Source: BBC.

still noticeable: 65% of those born in August get five GCSEs or equivalent at Grades A*–C, something achieved by 71% of September births. But August-born children take more risks when they are teenagers: more try smoking and cannabis, and they drink more alcohol. This is starting to look more familiar.

Compared with September children, we poor August things are less likely to go to university, and in particular to attend a top institution. Indeed, when the BBC made a Freedom-of-Information request to Oxford and Cambridge about undergraduates admitted in 2012, they found, for example, that 515 of those admitted were born in October, compared with 394 born in August. Overall, 0.7% of each year's 'cohort' get into Oxford or Cambridge, but Figure 52 shows this rises to 0.8% of those born in October, compared with 0.63% of August births. Getting into Oxbridge is only one rough indicator of educational achievement, but the 28% relative advantage of October births compared with August births is a remarkable finding.[28]

But there is a curious anomaly in university education. Multiple studies have shown that, of people who do get to university, those born in August tend to do *better* than those born in September, both in terms of having lower drop-out rates and getting better degrees.[29] A possible explanation is the 'selection' effect – the relatively younger people who do manage to get to university have had to make some more effort, and this is reflected in their performance once there. The same effect seems to operate in sports: 'older' football players (in their year group) were more likely than 'younger' players to drop out from elite levels of the German Bundesliga.[30] Presumably the relatively younger athletes have to work harder, or be better, to succeed in the face of their initial handicap, and if they do get through to an elite level, they reap the benefit of this harsher selection mechanism, in a bizarre sporting version of survival of the fittest.

I did get into Oxford, and have subsequently done OK in the academic world, utterly unaware I was labouring under an inborn handicap because of my November conception. I could almost, but not quite, feel sorry for myself.

12

PLEASURES AND PROBLEMS

In 2006 the World Health Organisation endorsed the following working definition: 'Sexual health is a state of physical, emotional, mental, and social well-being in relation to sexuality; it is not merely the absence of disease, dysfunction, or infirmity.'[1]

This is a radical statement. It says that sexual health is more than just being free of disease or physical problems: it's about feeling good about ourselves. And it doesn't say that a healthy sex life is necessarily a busy one: it's not like exhortations to do 20 minutes a day of exercise, or having five-a-day fruit and vegetables. Quality, rather than quantity, is what is important here.

The connection between sex and our general well-being is complex, and not easy to express in numbers. We naturally associate sex with physical pleasure, and we can measure how our bodies react to sexual arousal, and even try to work out whether sex makes us healthier. We can look at the stuff people buy to try to spice things up, and ask them how satisfied they are with their sex lives. It's not all perfect: many experience difficulties with sex, or 'dysfunctions', as they are known, and although rather few seek help, there's a small industry concerned with diagnosing and trying to treat sexual disorders.

So sex can bring pleasure and problems, and we might as well start with that perennial source of coarse humour – the size of the male 'organ', and its associated anxieties.

Does size matter?

~~~~~~

6: the average erect penis size, in inches

~~~~~~

When it comes to men, the main concerns of scientists have been ejaculation and erections – what could be called the nuts and bolts of sex. Of course, many men are concerned about these too, and we'll look at sexual dysfunction later. But for now here are a few odd – some very odd – statistics on the physiology of men's private parts.

Kinsey loved biological measurements: he asked about erections by indicating different angles with a pencil, and 15% to 20% of men reported being about 45° above the horizontal, and around 10% almost vertical, though there is an inevitable decline with age.[2] He also asked for the time of masturbation until orgasm: for the white male college-educated group the median was 2 minutes, while 1 in 20 said they were still going after 10 minutes. The median for females was 4 minutes.

Kinsey also asked men to estimate their penis size when erect, and then to measure it at home and send back the answer. The median estimate in white college men was 6 inches. When they measured themselves it was ... 6 inches. In spite of this apparently realistic insight, many men are unhappy about exposing themselves in shared changing rooms owing to a perception that their penis is small – this is officially known as 'small penis syndrome', or SPS. A fascinating review suggested that this was generally a misconception, and they contrasted this with having a genuine 'micro-penis', which they define as less than 7 cm (3 inches) stretched length.[3] The researchers considered it is important for doctors to have an idea of what is 'normal', as anxious men may not be realistic about their expectations.

They also concluded from multiple studies that there is no evidence for penis size depending on ethnicity, with typical lengths for a penis when flaccid to be 9–10 cm (around 4 inches), when stretched to be 12–13 cm (around 5 inches) and when erect to be 14–16 cm (around 6 inches), which is around the length of a British £20 note or a US dollar bill. There are a variety of treatments available to men who suffer from SPS – this may seem like something that may pop up in your spam filter, but studies claim that stretching devices were able to add 1 or 2 centimetres.

When it comes to satisfaction with size, the most cited statistics come from an internet survey run by *Elle* magazine in 2003, which attracted 52,031 responses, about equally split between men and women, aged on average in their mid-30s. Nearly half the men (45%) desired a larger penis, while 84% of women were satisfied with their partner's size, although this satisfaction fell to 32% in the group (only around 6% of women) who considered their partner's size as 'small'.[4] So if we believe this study, then size matters more to men than to women. But there are cases where it does seem to matter to women too.

It's not just men who get measured. A mean vaginal length of 6.3 cm was found from MRI scans on 28 women, with width ranging from 4 cm to 10 cm.[5] But this is a 'baseline' measurement: the size during intercourse is more related to the male measurements described above.

Ejaculation times

And so, after a suitable interval, we get to ejaculation. We've already noted that the average quantity of semen is around 3 ml, but what about the time taken for it to arrive once penetration occurs? This is officially known as the 'intravaginal ejaculatory latency time', or IELT. Many men feel their IELT is too short, but again, if we are to understand what is usually known as premature ejaculation, we need to know what is 'normal'. For obvious reasons this is quite tricky research to do, but, of course, it's been tried.

9: the median 'intravaginal ejaculatory latency time' in minutes, based on 914 couples (3*)

A huge European study used radio and newspaper advertisements to recruit over 1,100 couples from forty-four centres in France, Germany, Italy, Poland and the UK, and asked each couple to attempt intercourse at least twice a week for eight weeks, without trying any novel behaviours.[6] Couples were given a stop-watch, and asked to set it going on first vaginal penetration and stop it on ejaculation or withdrawal without ejaculation: presumably they were asked not to treat it as a race.

Around 200 men had already been assessed as experiencing premature ejaculation (PE), leaving 914 'normal' couples, and their median measured IELT was 9 minutes, but ranged up to an impressive 44 minutes. Men had been asked in advance how long they thought their time would be, and the median subjective assessment was 10 minutes, so they were fairly realistic. Except the Italian men, who had estimated a median of 15 minutes but came in at 9 minutes like everyone else. No comment.

Health benefits of sex

You are no doubt aware that lists are popular on the internet – in fact, I could probably think of at least five reasons for this – and it only takes a search (try 'health benefits sex') to find lists of 10, 11 and even 21 reasons why sex is good for you. The same claims, from pain relief to living longer, get trotted out time and time again, and illustrated with healthy-looking (but always partly clothed) young couples.

So let's look at some of these claims. We'll take for granted, without the need for rigorous scientific testing, that sex is generally pleasurable. And there is plenty of evidence that it helps you get to sleep, and can exercise a woman's pelvic floor muscles, providing a range of benefits from enhanced sensation to the prevention of 'leakage and incontinence' and 'post-menopausal vaginal atrophy' in older women. Women have also found that sex can provide pain relief, although it might be difficult to separate psychological from hormonal effects.

But will sex help you to live longer? A much-cited study of men in Caerphilly, from a respectable team, estimated that an extra 100 orgasms a year was associated with a one-third reduction in the risk of dying each year – this would translate to about 3 years' extra life.[†][7] A later re-analysis of the same data suggested that more sex was associated with fewer heart attacks but possibly more strokes over 20 years.[8] Of course, correlation is not causation, and we've already seen that healthy people have more sex, so maybe the causation is in the other direction. The authors tried to allow for this, but clearly we need to look at some more specific potential benefits of sex.

---

85 Kcal: average energy expenditure by young people during sex

---

A recurrent media claim is the benefits of sex as an exercise regime, although other estimates have estimated the energy expenditure at a paltry 21 Kcal, or about the same as in a satsuma.[9] So what is the truth?

We've seen in Chapter 8 that laboratory studies of sexual

---

† The date of publication reveals this study was in the Christmas issue of the *British Medical Journal*, notorious for tongue-in-cheek articles.

activity can be rather clinical, but researchers in Montreal solved this by instructing 21 volunteer couples to have sex once a week for a month in their own homes.[10] The only condition was that they both needed to wear a portable SenseWear armband during sex, which measured their energy expenditure, and to compare their sexual activity with a 30-minute treadmill test.

The sex sessions lasted an average of 25 minutes, and the average energy use was 101 Kcal for men and 69 Kcal for women, but the range was 13 to 306 for men, and 12 to 164 for women, showing a range from extreme mellowness to athletic exuberance among the participants. For calibration, 85 Kcal is about what you get in a very small glass of wine, a chocolate biscuit or a bag of cheesy Wotsits.

The average intensity of sexual activity was around 6 METS (1 MET is about the intensity of watching television), which is the borderline of moderate/vigorous, roughly between cycling and jogging. But some went up to 9 METS – higher than doing push-ups. Overall, as a source of exercise, sex came in at about half as effective as 30 minutes on a treadmill, although the participants (with one notable exception) rated it as considerably more enjoyable. These were healthy young couples with an average age of 23, but for older people maybe those METS might be all you need to bring on a heart attack?

A review of fourteen studies concluded that the risk of a heart attack did go up by nearly three times in the hours after sex for older people, but it was still only the same as any other form of similar exercise, such as shovelling snow,[11] and the risk depended on fitness. A plausible estimate is that if 1,000 middle-aged people had sex for an hour a week for ten years, then there would be around one sudden death and two to three extra heart attacks. This may seem a risk worth taking, and it is interesting to note that the majority of the case reports of sudden death in the past have been men engaged in 'extramarital sexual activity, in most cases with a younger partner in an unfamiliar setting and/or after excessive food and alcohol consumption'.[12] I think we get the picture.

Considerably less dramatic than the possible risk of death is the suggestion that exposure to semen can lighten a woman's mood, based on the theory that it's got lots of great hormones, which could create a positive mental state. But how can you test this? How can you expose some women to semen and others not, while controlling for the amount of sex going on? The researchers who tested this had a brainwave, realising that condom use provides a sort of natural experiment: women who have sex with condoms are deprived of this (possibly) psychically nutritious substance.

A study analysed the condom use of 293 college females; non-users scored an average 8 on a standard scale of depression, while those who usually or always used a condom scored around 12, where a score of 10 or more indicates 'mild depression'.[13] So there is a correlation, but not a very conclusive one: think of all the ways this apparent anti-depressant property of semen could be explained by some other factor. Was the benefit really coming from oral contraceptives? Or because non-condom users were having more sex, or in more committed relationships? The authors tried to allow for all of these, but were forced to admit that the data were 'only suggestive'.

There is one possible benefit that never seems to get mentioned in all these lists. Two doctors reported on one of their patients who had hiccups continuously for four days after a steroid injection, preventing him from working or sleeping.[14] Perhaps surprisingly, he did finally manage to have sex, during which the hiccups 'continued throughout the sexual interlude' – there is no mention of how his partner felt about this. This went on 'up until the moment of ejaculation when they suddenly and completely ceased', presumably to both parties' considerable relief. The medical conclusion is that masturbation might be tried when trying to stop intractable hiccups: an inexpensive treatment with few side-effects.

So, on balance, sex seems not only pleasurable but good for you. But maybe it's become a bit routine, and could do with a little enhancement – a little bit of spice.

Spicing things up

5%: the proportion of female customers of Lovehoney.co.uk who buy a male sex toy

Before we get on to sexual problems, and how people have tried to deal with them, it's worth briefly looking at what people do to bring a little something into their activities. Personal grooming is beyond the scope of this book, although it may be worth mentioning the recent move towards removal of pubic hair by women, which could perhaps be considered a sexual enhancement – a recent US survey reported a wide range of practices and with a strong influence of age: for example, for 16–24s the majority (59%) were typically or sometimes hair-free, compared with only 11% in women over 50.[†15]

But the enthusiasm for sexual enhancements is mainly shown by the booming market in sex toys and other products. Data journalist Jon Millward analysed 1 million products bought from online retailer Lovehoney.co.uk ('the sexual happiness people') over five months and found the breakdown of purchases shown in Figure 53.[‡16] They are worth looking at in detail.

The popularity of items such as lingerie, and 'lubricants

---

†The survey was of 2,451 women aged between 16 and 68, recruited by email in 2008 from various organisations to take part in an internet-based survey on lubricants. I would consider this 2* at best.
‡See jonmillward.com for revelatory details of some of the implausible products on sale.

Figure 53: **Breakdown of 1 million products sold by Lovehoney.co.uk over five months**

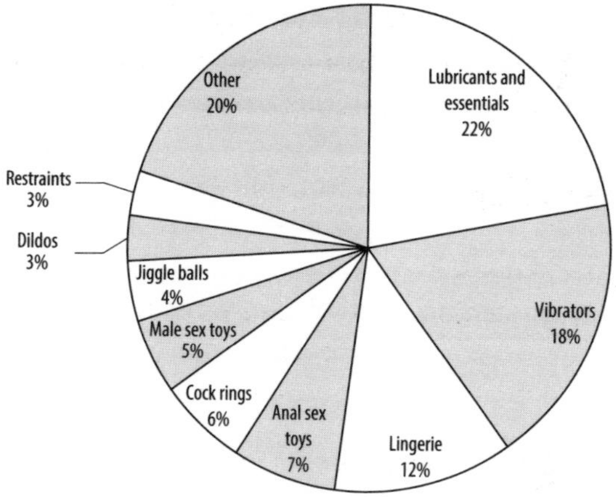

- Other 20%
- Lubricants and essentials 22%
- Restraints 3%
- Dildos 3%
- Jiggle balls 4%
- Male sex toys 5%
- Cock rings 6%
- Anal sex toys 7%
- Lingerie 12%
- Vibrators 18%

Source: Jon Millward.

and essentials', is fairly predictable, and we've seen that vibrators have been around for over a century. But the popularity of 'jiggle balls', for women to pop inside themselves as they go about their daily activities, most likely derives from *Fifty Shades of Grey*, while the sales of anal sex toys and 'restraints' reflect the increasing prevalence of the activities covered in Chapter 4.

These items make popular gifts: a third of male customers buy a vibrator, and 45% of corsets are bought by men. Presumably not all for themselves.

Satisfaction with sex

9%: of women who are not sexually active, the proportion who say they are distressed or worried about it (3*)

The WHO statement at the start of this chapter makes it clear that sexual health is not simply a lack of problems: it is a condition of well-being that has traditionally been treated under the label 'sexual satisfaction'.

We saw in Chapter 1 that 84% of Shere Hite's sample of US women in the 1970s were 'emotionally unsatisfied with their relationships'. But this rather dubious 1* statistic was contradicted by the 1992 US National Health and Social Life survey (NHSLS), which found that 88% of married people were 'very or extremely' physically satisfied, while 85% were emotionally satisfied, although there was some decline in satisfaction with longer marriages, even allowing for age.[17] It's a standard claim, reinforced by studies such as Hite's, that married men are more satisfied with their sex lives than women. The Irish sex survey provided some support for this: 82% of men said their sex lives were extremely or very pleasurable, compared with 77% of women.[18] But a recent Spanish survey found that 95% of both men and women said they were 'very or quite' satisfied with their sexual life.[19]

These rather positive images from Europe do not necessarily hold elsewhere. The Global Study of Sexual Attitudes and Behaviors (GSSAB)[†] asked 27,500 people from 29 countries

---

† The GSSAB was carried out by a respectable team, including Edward Laumann from the University of Chicago, who carried out the NHSLS survey in 1992 and the NHSAP survey in 2005. A mix of methods were used: random-digit telephone dialling in Western countries, door-to-door interviews in the Middle East and South Africa, by mail in Japan, and, most remarkably, in other Asian countries people were stopped in the street and asked to self-complete the questionnaire. 145,000 eligible people were asked to participate during 2001 and 2002; over 97,000 refused immediately, another 22,000 got fed up and interrupted the interview, leaving only 27,500 (19%) who completed. The response rate was highest (30%) in countries where the subject was directly approached on the street – it is too easy to put a phone down, and this echoes comments from the Natsal team about the importance of preparing possible participants. The researchers said the response rate was 57% if we ignore those refusing immediately without even knowing what the survey was about, and say the disease incidence found in the survey matches

Figure 54: **Attitudes to sex reported by women and men in three 'clusters' of countries**

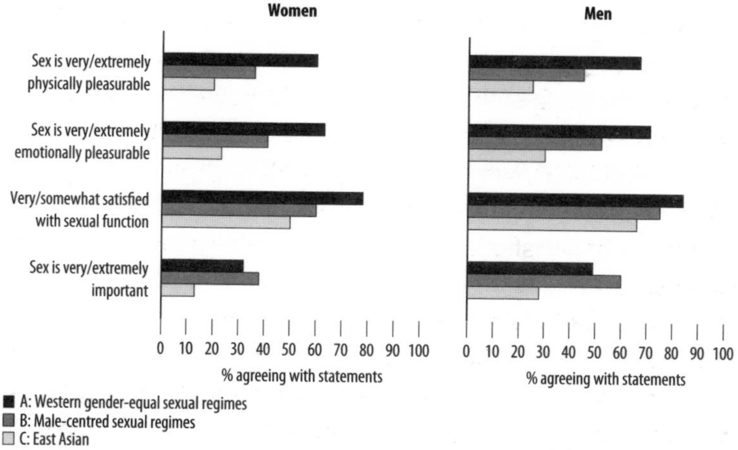

A: Western gender-equal sexual regimes
B: Male-centred sexual regimes
C: East Asian

Source: Global Study of Sexual Attitudes and Behaviors.

about their satisfaction opinions with their sex lives, and then used a statistical technique called 'cluster analysis' to find groups of countries that had similar responses.[20] They found three broad clusters of countries, which I have labelled in Figure 54. They termed the mainly Western 'A' countries as 'gender-equal sexual regimes', 'B' countries as 'male-centred sexual regimes', which were Islamic countries and Brazil, Italy and Korea, while 'C' countries were East Asian countries such as China, Indonesia and Japan.[†]

Regardless of context, women expressed less satisfaction than men – every bar on the left is shorter than the bar on the right. 'A', mainly European, countries are characterised by fairly high levels of satisfaction with sexual function, but 'B'

---

what is known from national data.

† 'A' countries: Australia, Austria, Belgium, Canada, France, Mexico, New Zealand, South Africa, Spain, Sweden Germany, UK, USA. 'B' countries: Algeria, Brazil, Egypt, Israel, Italy, Korea, Malaysia, Morocco, Philippines, Singapore, Turkey. 'C' countries: China, Indonesia, Japan, Taiwan, Thailand.

countries are distinguished by seeing sex as relatively important, particularly for men, but with lower levels of satisfaction, particularly for women. In 'C' countries there are very
low levels of satisfaction and sexual pleasure, and sex is not
seen as very important, particularly by women.

These patterns seem to reinforce national stereotypes to
some extent, but it is impossible to know how much of the
variation represents linguistic difficulties in trying to use
identical wording in the questions, and cultural distinctions
in how people are willing to respond to such questions, particularly when many are asked on the street.

In contrast, we should have a higher level of trust in the
responses to Natsal's crucial questions: are you dissatisfied
with your sex life, and are you distressed or worried about
it? This was asked of everyone, whether or not they reported
being sexually active in the previous year. The results, shown
in Table 3, deserve careful consideration.

Table 3: **Percentages of sexually active and not sexually active
women and men expressing agreement with statements**

Sexually active (partner in last year)	Women	Men
Dissatisfied with sex life	11%	15%
Distressed or worried about sex life	10%	9%
Not sexually active (no partner in last year)	Women	Men
Dissatisfied with sex life	22%	32%
Distressed or worried about sex life	9%	15%

Source: Natsal-3, 2010.

Only around one in seven sexually active people reported
being dissatisfied and around one in ten reported being distressed about their sex life, with little difference between
women and men. The responses from those who had not had
a partner in the past year are different, but not dramatically so:
there is somewhat more dissatisfaction, with a third of men
and a over a fifth of women dissatisfied, but still a low level of
distress. In fact, for women, the proportion who are distressed

or worried about their sex life is essentially the same, whether or not they are sexually active.

So only a small proportion of those who are not sexually active are distressed about it. This 'mustn't grumble' attitude might be considered either as British stoicism or a reasonable perspective on life's priorities. Or both.

How many people feel they have sexual problems?

34%: the proportion of women interviewed in Natsal-3 reporting a lack of interest in sex lasting at least 3 months in the preceding year (3*)

In spite of the fairly low levels of dissatisfaction and distress that were found by Natsal, this does not mean that people are not having difficulties with their sex life. We have to assume that it has always been so throughout history, even though little evidence has been left, except the tradition of aphrodisiacs and aids to masculine potence. To generalise enormously, the archetypal view of sexual problems has been that women's sexual dysfunction has stemmed from lack of interest or enjoyment, while men have wanted to do it but not been able to, or have got it over with too quickly.

Natsal-3 gives us an opportunity to investigate these stereotypes.[21] Sexually active people (those reporting one or more partners in the last year) were asked a series of specific questions about sexual problems that had lasted at least three months, with the responses shown in Figure 55.

Overall, the data provide moderate support for the stereotypes: the most salient feature of Figure 55 is the third of the women reporting a lack of interest in sex – note that lack of interest is not necessarily something 'wrong', but these respondents clearly experienced it as a problem. This pattern

Figure 55: **Proportions of men and women experiencing sexual difficulties**

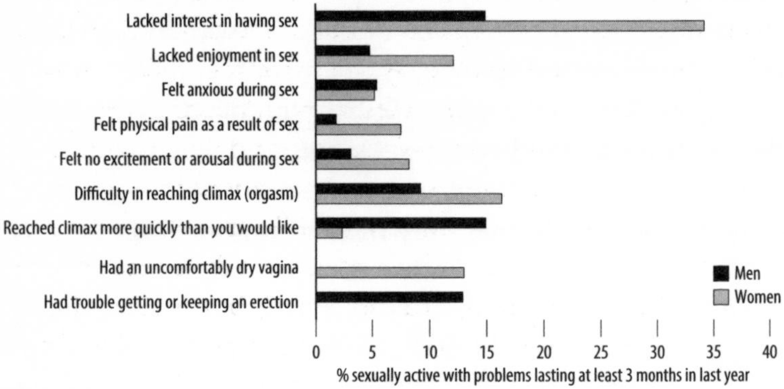

Lacked interest in having sex
Lacked enjoyment in sex
Felt anxious during sex
Felt physical pain as a result of sex
Felt no excitement or arousal during sex
Difficulty in reaching climax (orgasm)
Reached climax more quickly than you would like
Had an uncomfortably dry vagina
Had trouble getting or keeping an erection

■ Men
▨ Women

0   5   10   15   20   25   30   35   40
% sexually active with problems lasting at least 3 months in last year

Source: Natsal-3, 2010.

extends to around 1 in 12 women feeling no excitement or arousal during sex.

16: the number of Clelia Mosher's sample of 45 Victorian women who said they always or usually experienced a 'venereal orgasm'

Figure 55 shows that about one in six women in the Natsal survey reported that she 'Did not reach a climax (experience an orgasm) or took a long time to reach a climax despite feeling excited/aroused.' The female orgasm has been a topic of statistical controversy ever since Freud declared that only orgasms from male penetration were mature and healthy, whereas those from clitoral stimulation from, say, masturbation were a sign of immature development. Masters and Johnson counter-argued there was no essential difference in the physiological responses, and in the 1970s Shere Hite, like Kinsey, focused on orgasm as the defining aspect of sexual expression and fulfilment. She had no gradual lead-in: her second question was 'Do you have orgasms? If not, what do you think would contribute to your having them?' Her much-quoted (1*) statistic is that 30% have orgasms through penetrative sex alone, which means that 70% don't. In contrast, 82% of her sample masturbated, during which nearly all claimed to have an orgasm.[122] But in the 1992 US NHSLS, 29% of women reported they always had an orgasm during sex, 41% usually, 21% sometimes, 8% rarely or never.[124]

Men have difficulties too. Natsal-3 found that around 15% reported reaching a climax too soon – otherwise known as early, or premature, ejaculation – while trouble getting or keeping an erection, also known as erectile dysfunction, was reported by 13% overall, around 1 in 8, but with a strong correlation with age: 8% of those around 20 had that difficulty, increasing to 30% of those around 70. It's worth noting that the proportion of men reporting lack of interest, enjoyment, excitement and arousal is still around half that of women, with 'lack of interest' matching the rates of premature ejaculation.

22%: the proportion of women in Natsal-3 reporting two or more sexual problems lasting at least three months

---

Figure 56 shows the proportions of sexually active men and women of different ages reporting either at least one

---

† It's been suggested since 1924 that the ability to have an orgasm from penetrative sex alone is related to having a shorter distance between the clitoris and the vagina, with one inch quoted as a critical threshold. This has become a statistical issue, with recent re-analysis of data from 1924 and 1940 claiming to find this relationship.[23] I defer to Mary Roach's analysis of female sexual response in *Bonk*.

‡ It's worth noting that a century previously Clelia Mosher had asked her respondents 'Do you always have a venereal orgasm?', and got the response 'always' or 'usually' from 16 out of 45 women, while a further 18 confirmed occasions it had occurred. The wording of her question suggests what answer she expected, which could be surprising to those steeped in a traditional view of Victorian sexuality.

Figure 56: **Proportions of sexually active men and women reporting problems**

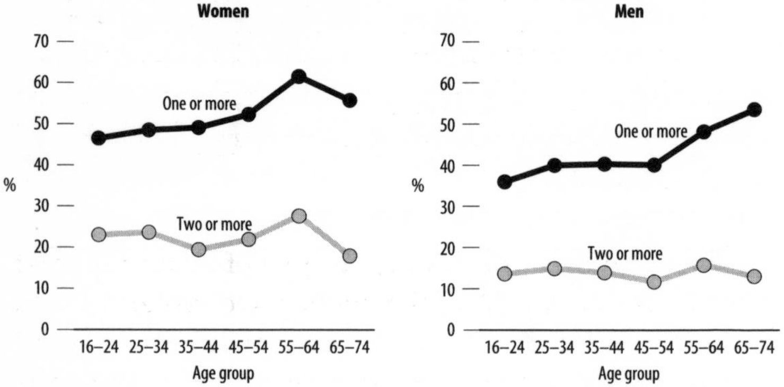

Source: Natsal-3, 2010.

or at least two of the problems listed in Figure 55. Overall, around 50% of women and 40% of men report 'one or more problems' (dark line), with the rates increasing with age. But the proportions seem high even for 16- to 24-year-olds: you are definitely not alone if you are young and your sex life is not perfect. Around 22% of women and 15% of men report 'two or more problems' (light line), and perhaps the most remarkable feature of the graph is the way the light lines are essentially horizontal: the incidence of multiple problems is similar for younger and older people.

Many people report physical sexual problems, but, as most sex occurs in a relationship, Natsal-3 also asked how they felt about their partner. The responses are shown in Table 4, and show remarkable similarity between genders: in around one in every four couples one partner is more interested in sex than the other, although, seen differently, this means that 75% of people feel their partner has around the same level of interest. Around 15% report a partner with sexual difficulties, but very few people do not feel emotionally close to their partner.

Table 4: **Proportion of people aged 16–74 expressing agreement with statements: Natsal-3, 2010**

	Women	Men
Partner does not share same interest level in sex	28%	23%
Partner has had sexual difficulties in last year	16%	17%
Partner does not share same sexual likes and dislikes	7%	10%
Hardly ever feel emotionally close to partner during sex	2%	1%

Source: Natsal-3, 2010.

As we have noted previously, in spite of all the reported problems, few seek help: of those who were sexually active, 17% of women and 14% of men sought help or medical advice from any source in the preceding year. And the percentage was higher in younger groups, in spite of experiencing fewer problems, again reflecting the uncomplaining attitude of older people about their sex lives.

The Natsal team have developed a multi-dimensional sexual function scoring method – the Natsal-SF – which combines all the questions discussed above. This rather predictably shows reduced sexual function with increasing age. If we just consider effects within each age group, Natsal found no relation of sexual function to economic deprivation, but it was worse for those who were unemployed, depressed or in poorer health. For those in a relationship, sexual function was lower if they were not happy, didn't talk about sex, had sex fewer than four times in the last four weeks, had been diagnosed with a sexually transmitted infection, reported many sexual partners in their lifetime and had experienced sex against their will.

These conclusions may not seem surprising, but, like so many of the associations discussed in this book, it is not clear what causes what. Nor can such a simple chain of causation ever be established in a complex and interconnected life in which sex is just one – albeit reasonably important – factor.

14%: the proportion of Americans aged between 55
and 85 who were using medicine or supplements
to improve their sexual function

The British are not alone in experiencing sexual problems. The US National Social Life, Health and Aging Project (NSHAP) studied the sex lives of older people between 57 and 85, and found that half of the sexually active men and women had 'at least one bothersome sexual problem', mostly reporting very similar problems to those found in Natsal-3, such as low desire, erectile dysfunction and inability to climax.[25]

On the whole, the British seem largely to resign themselves to their sexual problems, but Americans appear more likely to make a fuss. The US survey estimated that 38% of men and 22% of women aged between 55 and 85 had seen a doctor about sexual problems since they were 50.[26] One in seven was using medicine or supplements to improve their sexual function – presumably Viagra or something similar.

Sexual dysfunction exists all around the world, although it is challenging to produce comparable data. The Global Study of Sexual Attitudes and Behaviors (GSSAB), which collected the sexual satisfaction data discussed previously, was sponsored by Pfizer, makers of Viagra, and so it's no surprise that it also focused on sexual dysfunction among middle-aged and older adults worldwide.[27] They reported that the most common dysfunctions were the usual suspects: men reported early ejaculation (14% over all 29 countries) and erectile difficulties (10%), while among women the problems were lack of sexual interest (21%), lack of vaginal lubrication (16%) and inability to reach orgasm (16%).

Northern Europe consistently had the lowest rate of dysfunction, with 23% of men and 31% women reporting at least one problem, rather lower than Natsal-3 found: South-East Asia had the highest rate, with 44% and 55% of women with at least one dysfunction. This is not a hugely reliable survey, 2* at best, and there may be many cultural differences in how people respond to such questions. But put together with the data on satisfaction, a rather bleak picture of South-East Asian self-perceived sexuality is suggested. It is unclear whether this correlates with the market in rhinoceros horn and other 'remedies' to enhance sexual vitality.

In general, this whole area of research seems rather a mess, with little standardisation of definitions or subjects, so that different studies get wildly different answers. Nevertheless a recent review of the international literature concluded that around 40 to 45% of adult women and 20 to 30% of adult men had 'at least one manifest sexual dysfunction', and wherever there are medical problems there will be somebody trying to develop and sell treatments.[28] But deciding when a sexual 'problem' can be labelled as a 'disease' has a long and controversial history.

What is a sexual disorder?

26: the number of years that Hypoactive Sexual Desire Disorder existed as a diagnosis

The Diagnostic and Statistical Manual (DSM) published by the American Psychiatric Association (APA) is the bible of mental ill-health. It lays down what is a mental disorder and what is not, and, if a diagnosis is in there, how to decide if someone has it. The problem is that, in the area of sexuality, diseases come and go as attitudes change.

For example, from the first DSM-1, in 1952, homosexuality was listed as a sociopathic personality disturbance, following the Freudian rather than the biological view. It stayed there until, after much campaigning, it was downgraded in the 1974 edition of DSM-2 to a 'sexual orientation disturbance' and then was further refined to focus on cases in which distress was caused by conflicts about sexual orientation.

When it comes to sexual dysfunction, early DSMs included 'frigidity' for women and 'impotence' for men, but by 1987 DSM-3 had identified Hypoactive Sexual Desire Disorder (HSDD, focusing on lack of interest in sex), and Sexual Aversion Disorder (SAD, a chronic fear of sex). Numerous studies have been done trying to estimate how many people have these conditions.

But now those diagnoses no longer exist, so it's back to the drawing board for the statistics. In DSM-5 in 2013 sexual dysfunctions for women now include a combined female Sexual Interest/Desire Disorder (SIADD) and female orgasmic disorder; for men there is erectile disorder (still there, although now generally considered a medical issue), male Hypoactive Sexual Desire Disorder, premature ejaculation and delayed ejaculation.[29] But in response to the accusation that normal variation in desire was being classified as a 'disorder', it's been made more difficult to get these labels, needing at least six months with the condition and 'significant distress'. A 'hypersexual disorder' had been proposed but was voted down, so there is no official medical diagnosis of sex addiction.

If people can be given formal medical diagnoses, it means that treatments, if approved by regulatory authorities, will find a lucrative market. But not all treatments for sexual dysfunctions have waited for official approval.

Can we treat sexual disorders?

60%: the proportion of Viagra used in the Netherlands that is illegally obtained, estimated from sampling sewage

The long and dubious search for treatments for men who

wished to feel more potent was revitalised by the discovery of hormones. John 'Goat-gland' Brinkley travelled the USA in the 1920s, promising to restore sexual energy by transplanting goats' testicles into men – his radio broadcasts made him notorious, and he nearly got elected as Governor of Kansas in 1932, before finally going bankrupt amid numerous lawsuits over deaths of his patients. Eugen Steinach, in contrast, was no quack like Brinkley – he was head of Vienna's Biological Institute when, in 1912, he started conducting partial or full vasectomies in the belief that this would enhance virility by encouraging the production of testosterone rather than wasting time on sperm.†

A rather more reliable treatment was discovered by accident when a drug called sildenafil citrate was being tested on older men with angina, who reported an unexpected and welcome side-effect. When it was marketed as Viagra and introduced into the National Health Service in 1999, access was deliberately restricted to 'erectile dysfunction due to a medical problem' – this was both to restrict demand and to counter accusations that the NHS was sponsoring lust. One treatment a week was considered adequate, and it was optimistically felt this would keep treatments to around £10 million a year, but by 2012 there were 1,200,000 prescriptions for Viagra costing around £30 each, totalling £40 million.[31]

Despite the fact that the cost for generic sildenafil has come down over 90% since Viagra lost its patent in 2013, and you can now buy at least the first dose over the counter at a pharmacist, there has always been a thriving black market in bootleg potency drugs. This makes it difficult to judge

---

†Men in search of lost youth queued up for the Steinach operation. Freud was secretly Steinached, and so was the Irish poet W. B. Yeats, with the aim of pleasing his young wife. Another recipient, 70-year-old Albert Wilson, was billed to deliver a lecture in July 1921 in the Albert Hall on 'How I Was Made 20 Years Younger' – unfortunately he dropped dead the day beforehand.[30]

how much is being taken in any community, although Dutch investigators came up with an imaginative way of working out the total amount of Viagra consumed in Amsterdam, Eindhoven and Utrecht – they sampled the sewage coming from these cities for a week. From the metabolites they could estimate the total amount consumed, and then work out that only around 40% of this was from legal prescriptions, meaning that the majority, 60%, was obtained illegally.[32]

Sex is incredibly complex. The way that sex becomes less frequent with a longer relationship shows that sexual desire and activity do not just result from a biological imperative – they are largely driven by personal relationships and surrounding culture. Nevertheless the extraordinary success of Viagra – it even keeps cut flowers erect for longer[33] – has shown that what was once considered largely a mental disorder could be effectively treated with a pill. That's fine for men, but despite the clearly enormous market, and large investments by pharmaceutical companies, there is no equivalent of Viagra for women. Sex defies simple categorisation into biological or psychological causes.

Finally, although these numbers are informative, this is, of course, not a therapy book – if you feel you are experiencing some of the problems discussed above, stop reading and get in touch with professional services.

The WHO statement is clear: sexual health is more than simply a lack of dysfunction. It's a laudable ambition to expect your sex life to be associated with 'physical, emotional, mental, and social well-being', but it's also clear that this is not easy to achieve: many people, at all different stages of life, do have problems. One of the most remarkable findings from Natsal-3 is that those with problems tend not to be very distressed about them, and even fewer seek professional help. This attitude could be considered as showing admirable realism about what a 'normal' sex life entails, but could also mean that some remediable conditions are going untreated.

Another powerful result was that only a minority of those who were not sexually active said they were dissatisfied, and very few reported distress. Kaye Wellings summarised the changing attitudes thus: 'There used to be an imperative not to have sex, and then this switched full-swing to an ortho-doxy to have much as sex as you can, and if not there must be something wrong with you. But both of these attitudes do a disservice in terms of choosing the lifestyle you want.'

13

SEX, MEDIA AND TECHNOLOGY

There's always been concern about sex in the media. The last century featured a series of legal battles to make books previously banned on grounds of obscenity publicly available: the famous comment made by the Prosecuting Counsel in the trial of D. H. Lawrence's *Lady Chatterley's Lover* for obscenity – 'Is this the sort of book you would even wish your wife or your servants to read?' – was only made in 1960, even if it sounds like something out of the Victorian era. What would he, or his wife, or even his servants, have made of *Fifty Shades*?

Then the attention moved to television, and to sex in advertising, and it now focuses primarily on online and smartphone activity, whether pornography, sexting or social networks. There is particular concern about the influence on young people's attitudes and behaviour, and their early sexualisation. It's a persuasive theory – how could anyone, let alone young people, who are still forming their own sense of self and learning to handle their sexuality, not be influenced by the record levels of sexual content in the media that surrounds them? But we'll struggle to find convincing evidence that the media have a direct effect on sexual behaviour, although find plenty of evidence of influence the other way: studies of sexual behaviour clearly encourage bad coverage by the media.

Is there more sex in the media?

> **78%**: the proportion of advertisements in a sample of
> men's magazines that featured sexualised females

Sex sells. Possibly. Ever since 1885, when W. Duke and Son boosted their sales by introducing cigarette trading cards featuring alluring pictures of (heavily clothed) actresses, advertisers have been trying to gauge the appropriate level of sexual content. Companies such as Calvin Klein became notorious for overt eroticism, and there is a strong impression among consumers that advertising content has become more sexual in general.

To check this, a study looked at around a thousand full-page ads in six US consumer magazines in 1983, and compared them with 2003 to see what happened over those twenty years; they looked at *Cosmopolitan* and *Red* for women, *Esquire* and *Playboy* for men, and *Newsweek* and *Time* for everybody else.[1] In advertisements featuring both men and women the models got a lot closer to each other: intimate or very intimate contact went up from 20% to 47% overall. There was also an increase in the amount of flesh on display: women models wearing suggestive clothing, partially clad or nude, went up from 28% to 49% overall. But the rise was even greater in men's magazines, from 30% to 78%: that means that in 2003 nearly four in five ads in men's magazines featuring a sexualised female. But it's not just the men's magazines: the percentage of men's models in female magazines that were suggestive, partially clad or nude went up from 9% to 24%.

But how do people respond to all this extra sexual content? There is strong evidence that sex attracts attention: after all, why did you pick this book up? But there is a difference between looking at something and shelling out to own it, and accordingly advertisers are interested in what will actually make you buy something, and so carry out a vast amount of consumer testing. Most of this is kept secret, but some 'laboratory' experiments are carried out by researchers: for example, 242 students were randomly allocated either to see an 'explicit' advert for a deodorant or a 'mild' advert for a perfume,[†] and asked their attitude towards the ad, brand and whether they would consider buying the stuff. They found that explicit nudity – a bare bottom – was off-putting for both males and females, and that mild intensity was preferred.[2] The lesson seems to be that sex helps draw attention, makes the ad more memorable but can distract from the brand itself. It is notable that modern, vastly expensive, car adverts do not use sex, except possibly in a more subconscious manner.[‡]

Does sex on television influence young people's behaviour?

2%: the proportion of people who say they would like to see more sex on television

Older readers may remember Mary Whitehouse's National Viewers' and Listeners' Association, which campaigned for

---

† The 'explicit' featured a young, attractive and fully nude female shown from behind walking through a busy café district with the line 'feel confident, go naked', whereas the 'mild' ad just had a young couple sitting on a motorcycle, with the male leaning in to kiss the female on the neck, although his hand was on her upper thigh.
‡ A example that is sometimes held up as a successful use of sex was the US 'clean your balls' campaign for Axe shower gel: a girl with an immaculate plummy English accent asks 'How can guys clean their balls so they're more enjoyable to play with?' and later fondles two gold balls in her hand. But this is intended to be funny, at least to people who share my schoolboy humour, rather than being overtly sexual.[3]

less sex and violence on television.[†] Mrs Whitehouse was replaced by an official counterpart: opinions about sex on UK television are now monitored by the Ofcom media tracker of around 1,800 adults, and in 2001 nearly half (44%) of adults felt there was 'too much' sex on TV.[4] By 2013 only around one in four (26%) felt there was too much – it's unclear whether this reduction in offence is due to an increased appetite, or tolerance, for sexual content, or because television has cleaned up a bit. An enthusiastic 2% said there was not enough sex on television.

But does watching sexual content on television and other offline sources change the behaviour of young people? In spite of all this exposure, we've seen in Chapter 7 that the average age of first sex has been rising in the USA, but some prominent studies have nevertheless claimed a relationship. A much-cited paper unambiguously titled 'sexy media matter: exposure to sexual content in music, movies, television, and magazines predicts black and white adolescents' sexual behavior' reported a study of around 1,000 black and white teens, and found that a stronger 'sexual media diet' was associated with more sexual activity two years later in white teens, but not (in spite of the title) in black teens.[5]

But the problem, of course, is that the factors, such as early development of sexual interest, that might have led to more exposure to sexual media are precisely the ones that might lead to more sexual activity, and simple statistical regression analysis cannot adequately control for this. The data have recently been re-analysed using a technique called

---

[†] Mary Whitehouse once complained to the BBC Chairman Sir Michael Swann about a programme in which the artist Claes Oldenburg was heard to say: 'That's almost as big as my balls.' Sir Michael replied that Oldenburg had actually said 'that's almost as big as my ball', referring to a marble ball in his sculpture. The NVLA's judgement was also flawed when, in 1977, it gave Jimmy Savile an award for his 'wholesome family entertainment'.

'propensity score matching', which attempts to mimic the situation in which people were randomly allocated to watch sexual media or not.[†] They found, when analysed using these more appropriate methods, that the apparent effect of watching the media disappeared, and concluded that 'it is likely that the most important influences on adolescents' sexual behavior may be closer to home than to Hollywood'.[6] But by August 2014 the original study had 427 citations in the scientific literature, and the one that disputed its findings had only 25.

In any case, the heat has gone out of the debate about sex on television. Apart from a few entertainment programmes, young people hardly watch it any more – the median age of BBC1 viewers is now 59. Younger viewers use the internet for entertainment. And they also use it for porn.

How many people look at pornography?

3rd: the position of Egypt in the league table of countries accessing pornography website xnxx.com

In February 2009 the husband of Jacqui Smith, UK's first female Home Secretary, had to make an embarrassing public apology after he ordered two pay-per-view 'adult' movies that his wife subsequently claimed on expenses.[‡] It caused

---

[†] They put together matched groups of teens who are similar in the factors that predict watching sexual media, and then within each group look at the relationship between what they actually watched and their subsequent behaviour – this is the closest we can get to randomisation without randomising. The same technique was used by the team re-analysing the pledge data in Chapter 7.
[‡] Ms Smith had submitted her internet bill as 'wholly, exclusively and necessarily to enable me to stay overnight away from my main home'. But her internet package included TV, and the bill included

a big media fuss at the time, but was Ms Smith's husband doing anything very unusual? Trying to track down the statistics of pornography is particularly challenging, especially as there is no clear definition of what counts: US Supreme Court Justice Potter Stewart famously said of pornography, 'I know it when I see it.'

This is not as fatuous as it at first sounds – pornography is defined more in terms on the effect it has on us, than in terms of what it actually features. As Brooke Magnanti, aka Belle de Jour, says in her book *The Sex Myth*, erotica is what turns *you* on, pornography is what turns other people on.[7] Context and labelling make all the difference. Victorian gentlemen collected highly erotic art, which was, just about, socially permissible as long as it was based on classical themes: 'Leda and the Swan' was acceptable as a label, 'Naked Girl Being Seduced by Bird' would not have been.

Data about pornography tend to be used as weapons against it. The most voluminous statistics come from Covenant Eyes, which is undoubtedly an opponent: it sells 'Internet Accountability' software, starting at $9.99 per month, that automatically monitors usage and sends reports to your 'Accountability Partner', so that you know someone is checking what you watch.[8] We can take one statistic from Covenant Eyes and check its provenance: say, '68% of men and 18% of women said they used pornography at least once every week.' This turns out to be from a reasonably respectable 2005 survey of 688 Danish adults aged 18–30, and is probably a 3* statistic.[9] But this was Denmark, and a few years ago, in a specific, fairly young age group, and other studies come up with rather different figures. Sites such as Covenant Eyes use real statistics, but they are highly selected to make a point, and then assumed to apply generally.

---

two 18-rated films at £5 each. She lost her job on 5 June 2009 – it didn't help that she had claimed £116,000 expenses for a second home that turned out to be her main residence.

A good source of data that does not have an opinion to push is Amazon's Alexa.com, which lists the top 500 websites in the world. In September 2014 this featured xvideos.com at number 39, and Pornhub at 72. For comparison, Netflix was at 78 and vimeo.com at 100. In the UK, Xhamster.com was at 35 and xvideos at 39, so both have roughly the same number of hits as Tripadvisor at 36.

Alexa.com also report where the traffic for these sites came from: the top five countries for xvideos.com were the USA, Japan, India, Brazil and Mexico, while the top five for xnxx.com (globally number 84) were the USA, India, Egypt, Algeria and Italy. xnxx.com was the fourteenth most popular site in Egypt, while it came in tenth in Algeria, which speaks to the extraordinary reach of pornography in countries newly accessing the internet.

But how old are these viewers? A much-quoted YouGov poll carried out for Channel 4's programme *Sex Education* in 2008 claimed that 58% of all 14- to 17-year-olds have viewed pornography online, on mobile phones, in magazines, films or on TV, and more than a quarter of the boys looked at porn at least once a week, with 5% of them viewing it every day.[10] But, then again, this poll is based on a self-selected online panel, with no commitment to honesty, and is 2* at best.

Rather than relying on unreliable self-reports of behaviour, it would clearly be better to collect data on what the kids were actually looking at. Nielsen use a panel with 45,000 home computers that are wired to record all websites that are visited; tablets and smartphones are not included. In March 2014 the Authority for Television On Demand (ATVOD), the official regulator of TV on-demand services in the UK, headlined a report with the impressive statistic 'at least 44,000 primary school children accessed an adult website in one month alone', based on the Nielsen panel for December 2013. The 44,000 figure presumably comes from scaling up a very small number of observed visits, and Nielsen warned this did not 'meet minimum sample size standards', adding 'sample sizes for the youngest age groups visiting adult websites are relatively small and should be treated with caution'. But this heavily promoted 1* statistic served its purpose of attracting a lot of publicity to ATVOD's campaign to stop payments from the UK to foreign porn sites that did not provide effective age verification.

What are people looking at?

As with the rest of the book, I am trying to avoid the criminal and pathological, and so I am only dealing with what could be called 'mainstream' porn (although, again, this is difficult to define). If you want to know what goes on in films, and quite sensibly don't want actually to watch any, you could consult the convenient and exhaustive Internet Adult Film Database (IAFD).[11] This covers over 150,000, mainly American, films and over 130,000 'performers'. You can find, for example, that in 2011 India Summer (not her real name) made 78 films, while Rocco Reed (ditto) clocked up 133.

We've already considered data journalist Jon Millward's analysis of sex aids, and he's also looked at the IAFD. His sample of 10,000 porn stars showed the average female is 5'5", weighs 117 lb and started her career around the age of 22.[12] He has also analysed what each has done in their films: for example, 62% have had anal sex and 39% simultaneous anal and vaginal.

It can be a lengthy career: the ten most prolific men have been going for over 22 years, and have worked with more than 1,000 different women each. In contrast, Nina Hartley started in 1984 and has made a record-breaking 938 films, but only had sex with 199 different men, reflecting that women in the 'industry' tend to work with a limited group of male partners.

What effect does pornography have?

But despite the many thousands of people viewing porn every day, not everyone is a fan. Mainstream pornography has been described as 'a physical and unaffectionate game between uncommitted partners', full of negative gender stereotypes and exploiting both the participants and the viewers.[13] But does it actually harm the people who watch it? It's impossible here to cover the vast amount of work that has been done on the possible effects of pornography on people's attitudes and beliefs, their sexual behaviour and, in particular, sexual violence. But we can use a couple of examples to show the statistical problems in proving such a link.

As usual, we've got two main sources of statistics: laboratory experiments under controlled conditions, and surveys of groups of people. First, the experiments. These were fashionable in the 1980s: essentially, show people a lot of porn and see what effect it has on them. A classic study recruited 160 participants and randomly allocated them either to be exposed to six sessions of one-hour pornographic films or to watch current TV comedies.[14] At the end, those forced to watch the porn reported less satisfaction with their partners' physical appearance, sexual curiosity and sexual performance than those who saw the comedies. Unfortunately they did not measure their satisfaction at the start of the experiment, so a cynic would say that we don't know whether the porn made people less satisfied or whether watching re-runs of M*A*S*H and Cheers made people feel extra-affectionate and satisfied. Later experiments confirmed short-term effects, but, after all, these are rather unnatural situations with possibly unusual volunteers.

So the obvious step is to try and work out what is going on in the 'real' world using observational data. But then, of course, we have to just take what we are given – we can't make people change their lifestyle and start (or stop) looking at porn. For example, a recent survey of 782 US psychology students showed that use of 'sexually explicit materials' (SEM) is associated with higher-risk sexual behaviour and lower sexual satisfaction.[15] The author acknowledged that these were self-selected volunteers, but still managed to conclude that SEM use can play a 'significant role' in young adults' sexual development.

This may well be the case, but the data cannot prove it. Possibly it's the other way round: maybe people who do not feel sexually satisfied use more porn? And maybe there's some bias in the responses; those who are sensitive to social desirability might cover up both their porn-viewing habits or their risky sexual behaviour. As always, correlation does not mean causation.

The question that receives most attention in the media concerns the effect of internet pornography on adolescents. Dutch researchers Jochen Peter and Patti Valkenburg are leaders in this field: using an online panel set up by a survey institute that claims to be nationally representative, in one study they recruited around 2,300 young adults (18+) and adolescents (14–17) who supplied information on their risky sexual behaviour (sex without a condom) and their consumption of sexually explicit internet material (SEIM), and then repeated the exercise six months later – this was around two-thirds of those originally contacted.[16]

Rather surprisingly, only 30% of adolescents and young adults said they had watched SEIM in the six months before the first interview – much lower than the statistics usually quoted by sites such as Covenant Eyes: maybe the Dutch, who were ahead of the field, are just getting a bit bored with it all. And for the adolescents, there was no relationship between watching SEIM and subsequent risky sex, although there was a small association for adult males.

So nothing very dramatic, or conclusive. But yet again the investigators try to have it both ways: being scientifi-cally balanced and yet wanting to make bold claims that get attention. In this case, it meant that the Dutch authors (quite correctly) stated that 'our results must not be interpreted in the sense that exposure to SEIM caused, or led to, sexual risk behavior', but then couldn't resist titling the paper with the somewhat misleading 'The influence of sexually explicit internet material on sexual risk behavior'.

But just because we can be critical of attempts to 'prove' statistically that exposure to pornography has negative effects on adolescents, this does not imply we should be completely relaxed about it. Dolf Zillmann, who carried out the first experiments in the 1980s, said in 2000 that we do not need to show direct effects on violence or sexual coercion. Rather than focusing on the specific acts portrayed, 'should we not concentrate on the circumstances under which sexual access is sought and gained, and on the emotional consequences of sexual engagements?'[17] Zillmann concentrates on the formation of sexual 'callousness', of which pornography can be just one part, which easily leads to exploitation. And to sexting.

How popular is sexting?

25%: the proportion of 13- to 18-year-olds saying they had sent a sexual image or photo (2*)

Young people do not just have to be consumers of sexually explicit material: armed only with a smartphone, they can produce and distribute it themselves.

'Sexting' was first used in 2005 to describe sending sexually explicit messages or photos. A particular concern is the pressure on girls to take pictures of themselves, which can then get forwarded on to a wide audience, maybe in revenge for an ended relationship. But, as a teacher told me, intimate pictures of boys can also have a currency.

It's difficult to get even rough estimates of the prevalence of sexting, since the use of social media by young people changes very rapidly and the technology studied rapidly becomes out of date. While parents are worrying about Snapchat and Instagram, the kids will have moved on to something else.

This is another area where it is crucial who is asked, how they are asked and what is asked. Studies from the USA have come up with wildly differing conclusions, and some studies might just as well have set out to get deliberately low estimates. For example, in 2009, Pew Internet and American Life Project found only 4% of teens had sent sexually suggestive photos and only 15% had received them. This was based on 800 telephone interviews, but it turns out these were conducted using landlines in the teen's home, and after getting consent from their parents: it's not difficult to imagine that sexting adolescents would not be overly keen to volunteer, especially with their parents lurking in the next room pretending not to be listening.[18] And they had to make around 135,000 calls to get these 800 responses – these seem just 1* statistics.

A peer-reviewed survey published in 2011 also suggested that sexting was not such a big issue. But they again used landlines with a parent possibly present, and added to this by using an extreme definition of content: they found only 2.5% had appeared in or created nude or nearly nude pictures or videos, and only 1% in images that would be considered illegal child pornography.[19]

On the other hand, if you want a high figure, you should ask young people directly, or let them volunteer. For example, you could take the Kinsey approach of exhaustive sampling: go into a single private high school and ask all 600 students aged between 15 and 18. Nearly 20% said they had sent, and 40% received, a sexually explicit image.[†20] And in

---

† To someone who struggles to use smartphone keys, the most remarkable statistic was that the *average* number of daily text messages sent was 91, going up to a maximum of 300.

2008 the US Campaign to Prevent Teen and Unplanned Pregnancy, in partnership with Cosmogirl.com, used a market research panel and found 38% of teens aged 13–19 had sent a sexually suggestive message, and 19% sent nude or semi-nude pictures.[21]

In the UK, ChildLine reported a survey of 450 13- to 18-year-olds, in which 60% said they had been asked for a sexual image and 25% said they had sent an image or video.[22] But although the BBC reported that ChildLine 'spoke to 450 teenagers across the country', in fact, they filled up an online questionnaire using Survey Monkey.[23] ChildLine does acknowledge that this was a self-selecting group and may not be representative, but in fact it does not look far out.

It is striking that statistics are generally quoted for a wide age range, even though 13–17 covers a massive period of growing up, and what is disturbing in a 13-year-old might be part of a normal sex life for a 17-year-old. The National Society for the Prevention of Cruelty to Children (NSPCC) provides a sensibly uncertain estimate that between 15% and 40% of young people are involved in sexting,[24] but makes a strong distinction between different ages; Year 10s, aged around 15, tended to be more resilient, while Year 8s, aged around 13, were more worried, confused and in some cases upset by pressures they felt. The UK approach is generally through education and support, with excellent Childline and NSPCC resources, rather than bringing in the police.[25]

Sexting and pornography are all part of a general concern about the premature sexualisation of young people, which includes pressure to conform to airbrushed images in the media, with female stereotypes being typically thin, with beautiful hair and provocative behaviour, which is said to lead girls to have low self-esteem, an obsession with image and a wish to change their appearance. Boys have similar pressures, with some music videos and other media presenting men as sexually domineering and hyper-masculine.

251

Worthy and rather alarmist reports expound the dangers from exposure to sexualised media, in spite of the lack of direct evidence of harm.[26] But these are not just the opinions of over-anxious adults. The Institute for Public Policy Research think-tank surveyed a market research panel of 500 UK 18-year-olds, and found that at least two out of three felt that young people were too casual about sex and relationships, and that pornography was too easy for young people to see accidentally, that it could become addictive, that it could have a damaging impact on young people's views of sex or relationships and that it could lead to pressure on girls and boys to act a certain way.[27] These are 2* numbers from a self-selected group of 18-year-olds; 16% considered themselves gay/lesbian or bisexual, which seems high. But you may think they sound a sensible lot.

So young people themselves can see possible problems with an early exposure to sexualised media, but there is also remarkable resilience to influences that might appal older generations. Overall there does not seem to be a huge crisis. But don't expect that story in the media.

How is sex research reported in the media?

30%: the proportion of sexual encounters that take place before 16 (according to the *Mail Online*)

The media coverage of sex research has always, to put it mildly, tended towards the sensational, and in controversial areas media outlets pick and choose statistics that suit their argument (as I am sure I do as well).

Sometimes the coverage is just wrong. The Channel 4/ YouGov poll of adolescents in 2008 found that 30% of 15-year-olds were at least occasionally sexually active, just as Natsal did in 2010.[28] It also reported that, *of all those who were sexually active*, two-thirds had started their sexual activity before age 16.[29] *The Guardian* misunderstood what the 'two-thirds' referred to, and reported this as 'Two-thirds have had sex under the age of consent' – double the actual figure.[30]

If you think this is a bit innumerate, consider what happened while I was writing this book. Anne Johnson, from Natsal, and I gave a talk at the Cheltenham Science Festival about the statistics of sex, and the *Sunday Times* reported, correctly, that 'more people are having sex in their teens, roughly 30% before the age of 16'. But then the *Daily Mail* lifted this material into its own story. It made a number of errors, but the cracker was when the statement by the *Sunday Times* got turned into the remarkable headline: '30 per cent of total sexual encounters take place before 16'.[31] And this was attributed to me.[†]

This could be just an enthusiastic sub-editor, such as the one who produced the wonderful headline 'Over a third of bikers are killed on Kent roads', which was rapidly changed to the more reasonable 'Bikers involved in more than one third of serious Kent road accidents'. Note that these two errors are both of an identical logical nature – the so-called 'transposed conditional'. Here are some pairs of statements, the first of which is reasonable, the second in error:

- 30% of people, when they were aged under 16, had sex
- 30% of sex happens with people aged under 16

[†] A little reflection should show that the *Mail*'s statement is more than implausible. In Chapter 2 I estimated that opposite-sex couples in Britain have sex around 900,000,000 times a year. So if 30% of this really were happening to the under-16s, that would be about 300,000,000 times a year. There are about 1,500,000 14- and 15-year-olds – that's 750,000 potential couples – so they would all have to be having sex 400 times a year, which is more than once a day. No wonder they don't have time for homework. Or maybe this number is just ridiculous.

- One third of fatal accidents involved motorcyclists
- One third of motorcyclists have fatal accidents

- 90% of women with breast cancer get a positive mammography
- 90% of women with a positive mammography have breast cancer.[†]

In abstract terms, what happens is that the 'proportion of A that are also B', is reported as 'the proportion of B that are also A'. This is also known as the 'Prosecutor's Fallacy', as it is a mistake made in legal cases: it is extremely dangerous to mix up the statements

- The probability of the evidence, if the suspect is innocent, is 1 in 1,000,000
- The probability of the suspect being innocent, given this evidence, is 1 in 1,000,000

and yet this error in reasoning has implicitly happened repeatedly in legal cases. The essential problem of the *Daily Mail* is that the '30%' has been taken as a proportion of all sexual activity, rather than of people. This might be avoided by saying, for example, 'Out of every 100 people reaching 16, 30 have already had sex.'

Is this innumeracy or lack of logic? Perhaps neither: it's more an apparent inability to combine reason with numbers, some kind of mental paralysis that comes when confronted with numerical arguments that means that ordinary common sense is bypassed, especially when the topic is sex. One consequence of this inability to take a sensible critical attitude to numbers is that opinions are pushed to the extremes: numbers are to be either accepted, and even fetishised, as some sort of God-given truth or rejected out of hand as 'just statistics'. Possibly in the same breath.

[†] The current breast-screening leaflets point out that only around 25% of women with a positive mammography have breast cancer.

In this chapter we've seen many efforts to use statistics to 'prove' that the sexualised media, whether it's off- or online, causes harm, particularly to adolescents. The statistical claims can be pulled apart and their conclusions called into question, and so these no doubt worthy attempts seem doomed when the reasons for behaviour are so complex and we cannot randomly allocate people to be exposed, for example, to different amounts of porn. Beyond trying to describe accurately what is going on, statistical arguments are inevitably limited.

The one thing we can be sure about is that the use of technology will change our lives. A survey of 1,000 phone users claims that 9% of people have used their smartphones during sex,[†32] while the same technology now allows you to measure in precise numerical detail your sexual performance, including setting targets for 'thrusts per minute', all rather like a mechanical piston – Shere Hite, where are you?[33] And this world is changing so fast that any research into the effects of media and technology dates quickly. Apps such as Tinder and Grindr make it easier for people to meet up for casual encounters, and will no doubt have an effect on behaviour. They should also generate some extraordinary statistics.

Of course, there may be many developments in society that we don't like – the premature sexualisation of young people, exploitation of women in pornography, the portrayal of callous sexual relationships and so on. But rather than use spurious 'scientific' reasons for the harms that these media cause, perhaps we should be bold enough to say simply that these run counter to what we feel is appropriate. And try to make sure young people understand what is being pushed at them and, if they wish, feel confident in pushing it away.

†Online panel from Harris Interactive: 2* at best.

14

THE DARK SIDE: PROSTITUTION, THE POX AND HAVING SEX AGAINST YOUR WILL

How many prostitutes were there in London?

80,000: the bishop of Exeter's estimate of the number of prostitutes in London in the 1850s (o*)

One night in 1857 William Acton, whom we've already met discussing masturbation, continence and female sexual desire, walked home from the Opera in Haymarket to his home in Portland Place, a distance of less than a mile, during which he counted 185 prostitutes walking the streets.[†1]

This meticulous counting may seem odd, rather akin to Galton and his beauty survey, but there was a Victorian obsession with the statistics of prostitution. Back in 1796 Patrick Colquhoun, a police magistrate, had guessed there were 50,000 prostitutes in London.[2] These comprised 2,000 'well-educated women', perhaps 3,000 'above the rank of

---

†Throughout this chapter I will use the terms 'prostitute' or 'sex worker' according to the usage of the people whose work I am discussing. I shall take them as interchangeable and so do not include as 'sex workers' those, for example, who work on chatlines or as exotic dancers.

menial servants', 20,000 of the lower classes 'who live wholly by prostitution', and finally about 25,000 'who live partly by prostitution, including the multitudes of low females, who cohabit with labourers and others without matrimony'. The last category is an interesting example of why definitions need to be studied quite carefully: working-class women living with partners were considered prostitutes.

The population of London was then around 1,000,000, and so Colquhoun was categorising around 1 in 5 of all women between 15 and 40 as prostitutes. The nice round total of 50,000 got endlessly repeated, until it was upgraded to 80,000 by the bishop of Exeter in the 1850s, by pulling yet more numbers out of the air, and this figure continued to be used throughout the century.

But Acton, in his turgid report 'Prostitution Considered by its Moral, Social and Sanitary Aspects', pointed out that in 1857 the London police estimated there were only 8,600 prostitutes in total; 1,000 'well-dressed, living in brothels', 2,600 'well-dressed, walking the streets', and around 5,000 'low, infesting low neighbourhoods'. That's only a tenth of the bishop of Exeter's claim.

These statistical arguments raged on: in the classic study *London Labour and the London Poor* Henry Mayhew disputed the police figures as 'far from being even an approximate return of the number of loose women in the metropolis. It scarcely does more than record the circulating harlotry of the Haymarket and Regent Street',[3] which is not what Acton found on his statistical walk home.

An American visitor to London in the 1860s reported being accosted seventeen times between Haymarket and his hotel in the Strand, about once every fifty yards: a standard approach by a woman was to ask the time or to be helped across the road. But the trade rose and fell with the weather and the economy, and by the 1870s prostitutes had become fewer, less intrusive and better dressed, although in 1883 there were still reckoned to be about 500 in the Haymarket area, traditionally the epicentre of London prostitution.[4]

To the devoutly religious, a prostitute was a woman of unbridled passions, a temptress to be controlled: like Colquhoun, missionaries for 'fallen women' would lump together streetwalkers with working-class cohabitation. But William Acton rather boldly stood out by concluding that economic need, rather than sensuality, drove women into prostitution, and that it was an inevitable consequence of how society worked. His main conclusion was that prostitution should be regulated in order to prevent the spread of disease, a view that could be seen as part of the general shift in the 1800s from a moral to a 'scientific' basis for controlling behaviour. And, as we'll see, he was successful.

69%: the proportion of US men in the 1940s
who had visited a prostitute (Kinsey, 2*)

Victorian society had a particularly extreme version of the classic double standard in relation to male and female sexuality, in which women were supposed to remain pure while it was accepted that men could, and might even need to, play around, as is reflected in a passage from Acton's advice to newly married men:

> Again, it is a delusion under which many a previously incontinent man suffers, to suppose that in newly married life he will be required to treat his wife as he used to treat his mistresses. It is not so in the case of any modest English woman. He need not fear that his wife will require the excitement, or in any respect imitate the ways of a courtesan.[5]

This standard meant that the use of prostitutes used to be very high by modern Western standards. One of Kinsey's headline statistics was that 69% of white males in the USA in the 1940s had some experience with prostitutes, but with a big difference across educational backgrounds: 74% of men with only basic education had visited a prostitute by age 25 as a standard rite of passage, whereas only 54% of those who attended high school had done so, and 28% of college men. He estimated that around 4% of 'total male outlet' was with prostitutes, representing around five times a year for the average man.

Kinsey's statistics must be treated with great caution, but his claim that the older generation had around twice as much experience with prostitutes as did younger men seems plausible: he graphically describes previous generations visiting brothels for drinking, gambling and contacts often limited to five minutes or less with girls who 'expected neither nudity nor variety'. This change over generations was also noted in a survey that estimated that 26% of Norwegian men born between 1927 and 1934 had visited a prostitute, compared with only 6% of those born between 1975 and 1984.[6]

How many men pay for sex?

9%: the proportion of British men who report they have paid for sex at some point in their lives (Natsal-2, 2000)

Things have changed. In 2010 Natsal-3 estimated that 3.6% of men had paid for sex in the previous five years, around 1 in 30, but in the more sexually active 25–34 age group the proportion reached 5.4%, around 1 in 20, similar to the Norwegian survey mentioned previously. The Irish survey showed 6.4% of men had paid for sex in their lifetime, and 3.3% in the previous five years: again similar to Natsal-3.[7]

The proportion of men paying for sex in Britain grew between 1990 and 2000, from 2% to 4% of men aged 16 to

44 paying for sex over the previous five years.[8] The highest proportion was for single London men between 25 and 34, and there was no association with ethnicity or social class. 9% of men (1 in 11) reported paying for sex at some time in their life, and men who paid for sex reported more sexual partners and more sexually transmitted infections, although in the UK prostitutes are not the primary 'source' of infection: condoms are more likely to be worn than with non-paid partners. Very few women reported paying for sex in Natsal-3, around 1 in 1,000 overall – 8 women in total out of over 8,800 surveyed.

International surveys are generally based on asking a sample of men whether they have had sex with a sex worker in the last twelve months. The HIV/AIDS Survey Indicators Database collects together data from around the world and reports high rates in some South-East Asian countries: in 1999 over 62% of Cambodian military and police reported visiting a brothel in the previous year.[†9] Annual rates of paying for sex are also high in sub-Saharan Africa, with around 10% of men in Tanzania and Zambia, and this does not include customs such as providing gifts or household support in exchange for sex; some professions have a deservedly poor reputation, with 18% of Indian truck drivers in Tamil Nadu in 2000 reporting paying for sex in the past year, compared with 2% of factory workers in the same region.

Rather than interviewing men, some have tried to survey sex workers themselves, and that, as you can imagine, is challenging, particularly when selling sex is illegal, as in much of the USA. But there have been some resourceful attempts to get random samples of streetwalkers: interviewers in Los Angeles identified appropriate areas of the city, chose a random path and selected the first woman encountered who was judged likely to be a prostitute. Out of 1,629

---

† I would judge these as 2* statistics at best, but they still give a reasonable indication of rates.

women contacted, not all of whom would have been prostitutes, 998 were successfully interviewed on park benches or in fast food restaurants, laundromats, parking lots or the interviewer's car.[10] Not all potential contacts were followed through, as conditions were often considered unsafe; this survey required courage, both from the interviewers and from the respondents.

How big is the prostitution industry?

~~~~~~~~~~~~~~~~~~~~~~~~~~~~~~~~~~~~~~

£5.7 billion: the preliminary estimate by the
Office for National Statistics of the contribution of
prostitution to the UK economy in 2012 (1*)

~~~~~~~~~~~~~~~~~~~~~~~~~~~~~~~~~~~~~~

Suppose you want to know many fish there are in a pond. Here's a neat trick: catch 100, mark them and throw them back. Then wait for a few days for them to mix up nicely, and then catch another 100. If, for example, 20 of these new fish are marked, then you can estimate that there are 500 fish in the pond.[†]

The same 'capture-recapture' methods can be used to estimate how many prostitutes there are in a city. Colorado Springs Health Department knew of 247 prostitutes working between 1985 and 1988 – let's treat these as the 'marked' ones.[11] They compared their data with the police, who had 296 prostitutes on record over this period: let's think of these as the second batch 'caught'. There were 234 names in common, so that 234/296 = 79% of the police's 'catch' was 'marked'. This suggests that 247 is 79% of the total number

---

†Since if they are all mixed up, we can estimate that 20% of the fish in the pond are marked. We know that there are 100 marked fish, and so 100 is 20% of the total, which makes us think there are around 500 fish in the pond.

of prostitutes, which we conclude is around 312.[†] Of course, we've got to assume that everyone stood the same chance of being 'caught' – prostitutes that were good at keeping out of the way of both the police and the Health Department would be the equivalent of fish that hide at the bottom of the pond. They finally estimated 23 prostitutes per 100,000 population.

It may surprise you to know that the people who really want to know the numbers about prostitution in Britain work for the UK Treasury. Starting in September 2014, it has been mandated by the European Union that the trade in prostitution (and illegal drugs) must be included in the National Accounts and assessments of UK Gross Domestic Product (GDP). So the civil servants in the Office for National Statistics (ONS) have to produce official estimates of the current extent and cost of prostitution.[12]

Table 5: **Preliminary assumptions made by the Office for National Statistics in May 2014 concerning prostitution in the UK: all at 2009 levels**

Number of prostitutes in the UK	61,000
Clients per week	25
Weeks worked per year	52
Average price per visit	£67

Source: Office for National Statistics.

Preliminary judgements are shown in Table 5. These imply a total of 61,000 × 25 = 1½ million visits per week, or 80 million a year. There are some extra minor considerations of expenses and so on, but these assumptions alone give a total expenditure of 80,000,000 × £67 = £5.3 billion at 2009 prices (£5.7 billion at 2012 prices), which was the figure quoted by the ONS in May 2014.

But these estimates were immediately and vigorously attacked. Dr Brooke Magnanti, former sex worker and author

---

† Formally, 312 = 247 × 296/234.

of the Belle de Jour blog and books, thought that the £5.3 billion might even be ten times too high,[13] while the blog Tax Relief 4 Escorts, written by an accountant with a Cambridge mathematics degree, did a detailed critique of the figures, emphasising that while organising prostitution is illegal, selling sex is not: the UK tax authorities recognise prostitution as a 'trade or profession', and some sex workers were already paying tax and so were not part of the black economy, as Brooke Magnanti did when she was working in that role.[14]

Let's look at the numbers in Table 5. The total number of prostitutes was based on a 2004 London survey by the Poppy Project, a charity to help sex workers, who phoned escort agencies, massage parlours and saunas selling sex in London and asked about the numbers of women and their nationalities and ethnicities.[†15] There are a number of problems with using this survey as a basis for estimating numbers working in prostitution. First, it will have missed the informal and independent sector of both male and female workers, but it will also have over-counted through the practice of women working part-time at multiple establishments – maybe these would cancel each other out, maybe not. The second problem with the ONS assumptions is their scaling up the London estimates as if the same density of sex workers per 100,000 population existed throughout the UK. But Natsal and others have shown that a greater proportion of men pay for sex in London than elsewhere, and a small survey I did of the last 500 reviews on the prostitution review website PunterNet, 40% were from London, which has only 16% of the UK population.

---

† The ONS rounded the Poppy Project estimate to 7,000 off-street prostitutes in London in 2004, and then combined this with a 2004 Home Office estimate of 115 on-street prostitutes in London at any one time, which was then scaled up to give an estimated grand total of 58,000 prostitutes in the UK in 2004, or 61,000 in 2009 allowing for population growth.[16]

An alternative method, rather similar to the capture-recapture method used in Colorado Springs, was based on the UK Network of Sex Work Projects.[17] The average number of sex workers on the books of each project was estimated at 316, based on 38 of 54 projects that replied to a survey, with minimal double-counting; around a third were street-workers, and two-thirds working from premises, and around 9% were men. Assuming around half of sex workers were in touch with projects, and scaled up to the 153 services in the UK Network, this gives a total around 48,500, but the authors emphasise the uncertainty around this estimate.

Yet another survey was Project Acumen, conducted by the Association of Chief Police Officers (ACPO) in 2009, who tried to identify all businesses selling sex in each region by collecting advertisements, expert knowledge and cards in public telephone boxes (a practice already passing into a folk memory).[18] They estimated a total of 30,000 for England, which would scale to 36,000 for the UK, slightly lower than the estimate based on the UK Network of Sex Work Projects analysis, but which only included off-street female workers.[†]

But this is a rapidly changing industry. The internet has made it easy to start as an independent worker, and the website adultwork.com advertises over 30,000 individuals, including 11,000 men. Perhaps, although based on a rather inadequate method, the ONS figure of 61,000 in the UK is not too bad.

The next issue is the average cost per visit. The government statisticians estimated this from PunterNet.com,[19] a website that functions as a sort of Tripadvisor for British prostitutes,

---

†For example, ACPO identified 2,100 premises in London, phoned and visited a sample and estimated an average of 1.7 beds per establishment. Then, allowing for days off (2 per week for foreign workers, 3½ a week for UK), they estimated there would be 5,240 women involved in off-street prostitution in London, at the low end of the Poppy Project estimate.

hosting over 117,000 reviews, listing in meticulous detail how much has been paid, for how long and what precisely went on. The reviews are open for anyone to read and Jon Millward, already mentioned for his sex-aid and porn-star analysis, has turned his attention to PunterNet and examined 5,000 reviews: 500 from each of ten UK cities.[20] The average cost per hour was £103, and the average session lasted an hour and a quarter. The customers were mainly pleased with their experience: Jon noted that 87% would return and 92% would recommend the worker to other punters. Where there are criticisms, the worker can reply (often in forthright style).

Jon told me he thinks the PunterNet reviewers are fairly representative clients, but that PunterNet doesn't account at all for street prostitution and hardly at all for higher-class escorts. He thinks it's good that ONS are using an actual resource that, to some extent, reflects what men and women are getting up to and for how much. But it is somewhat ironic that a government agency is using a source that Harriet Harman, when Minister for Women and Equality in 2009, tried to shut down.[†21] Overall, considering the prices quoted on PunterNet and adultwork.com, my judgement is that the £67 average cost per visit quoted by the ONS may be a bit low.

Perhaps the most questionable figure is the workload: 25 clients each week, 52 weeks a year. This is based on Dutch experience, but in Britain there are great differences between the working practices in parlours, as escorts, as independents and on the streets. Many are part-time; some have families.

---

† PunterNet is hosted in California, and Harman unsuccessfully asked the Governor of California, Arnold Schwarzenegger, to close the site down. The resulting publicity generated a huge increase in calls. *The Independent* reported 'Laura Lee' saying 'A good review on a website like PunterNet results in a marked number of people requesting our services. It is to working girls what a Michelin star is to a restaurant.'

This workload, coming to 1,250 transactions per woman per year, seems implausibly high as an average.

There are other 'reality checks' that can be made on the ONS provisional figures. For example, it would mean that the average sex worker was turning over around £100,000 a year, which Tax Relief 4 Escorts said was not plausible, and they should know.

We can also look at this from the client's perspective. There are around 20 million men between 18 and 65 in the UK (taking an arbitrary upper limit), and around 2% of men report to Natsal that they paid for sex in the previous year. This may be an undercount, but also many will have bought sex in, say, Amsterdam or Bangkok, and so I would guess there are considerably fewer than 500,000 men buying sex each year in Britain. But the ONS estimate 1½ million transactions a week, so the average client is supposed to be buying sex at least three times a week. I am no expert on the behaviour of this subgroup, but this seems absurdly high: a study of men who pay for sex in Scotland found a mean of only five partners in a year.[22]

These apparently rather flawed preliminary estimates received substantial publicity when in October 2014 the European Union presented the UK with a bill for £1.7 billion in back-payments due to the UK economy doing better than predicted, partly due to the new back-dated contribution of prostitution. The ONS are revising the estimates: I bet they will come down substantially.

How many women are trafficked for sex?

25,000: the number of trafficked 'sex slaves' claimed by MP Denis MacShane in 2009 (0*)

Some 'massage parlour' websites now feature a rota of

women providing services and a price list.† But hanging over this increased normalisation of the sex industry is the accusation that many are working against their will, which brings us to the highly controversial issue of trafficking statistics. MP Denis MacShane claimed in Parliament in 2009 that 25,000 women had been trafficked into Britain as 'sex slaves',[23] while in the same year a *Guardian* comment piece claimed that 'In Britain, it is estimated that 80% of the 80,000 women in prostitution are foreign nationals, most of whom have been trafficked.'[24] The next day, Dr Belinda Brooks-Gordon of Birkbeck, University of London, who had worked on the UK Network of Sex Work Projects analysis, countered by arguing that inflated trafficking statistics are used by those aiming to abolish prostitution, who deliberately conflate legal sex work with illegal trafficking and abuse.[25]

It should be clear that this is a highly contested area, and establishing accurate figures is extremely difficult; there is frequent confusion between those coerced into prostitution and those being willingly smuggled. A frequently quoted 'Home Office' figure of 4,000 was acknowledged by the researchers to be a poor upper bound based on assuming most foreign sex workers working in flats had been trafficked.[26] Project Acumen, which we saw estimated around 30,000 women working off-street in England, also interviewed a sample of 254 women from 142 premises and concluded that 24 (around 10%) had been 'trafficked', although none had been kidnapped or imprisoned, and very few subjected to violence. These 24 women were scaled up to give an estimate of 2,600 trafficked women in England, a figure that gets many quotes. But it is unclear how the visited premises were chosen, and if they were selected as being thought to contain victims of trafficking, the statistic becomes highly unreliable.

There is now an official 'National Referral Mechanism'

† Gfemassage.co.uk currently advertises £45 for a 30-minute 'basic service', up to £140 for an hour in the VIP room.

that deals with potential victims of trafficking, which received over 580 referrals for sexual exploitation of adults in 2013. This was a steep rise on the year before, although there were twice this number of referrals for labour exploitation and sexual exploitation of minors: around a third of referrals were eventually formally recognised as 'victims of trafficking'.[27]

My personal judgement is that the number of victims of true coercion into prostitution is considerably lower than the 2,600 derived from Project Acumen. These serious crimes rightly arouse strong condemnation, but the poor statistics tend to be used as a more general weapon against the sex industry. Thinking back to the argument about the proportion of the population that is gay, it does not seem to be a coincidence that the most contested areas are those with the poorest statistics.

How has prostitution been 'controlled'?

2½ million: the number of signatures on the petition to repeal the Contagious Diseases Acts in the 1880s

The connection between prostitution, disease and the military has a long history. When Professor Falloppio developed the condom in the 1560s, he was concerned not with contraception, but rather with reducing the spread of syphilis – termed 'the French disease' because reputedly it was first spread by a group of French soldiers in 1494. He claimed to have tested his invention on 1,100 soldiers, none of whom caught the infection – such a suspiciously high success rate that I reckon this is o* data. By 1763 they were in use in the general population: James Boswell recorded how he had protected himself against disease by having sex 'in armour' with a girl on Westminster Bridge.

And well might he be careful, as disease was rampant, and Boswell himself died from kidney infections brought on by repeated gonorrhoea infections. William Acton estimated that in the 1830s and 1840s half the surgical outpatient attendances at one London hospital were for venereal diseases. The army was a particular concern: out of 97,700 British soldiers in 1860 there were 13,700 hospital admissions for primary venereal sores, 3,300 for secondary syphilis and 13,000 for gonorrhoea.[28] Something had to be done and Acton, the medical journal *The Lancet* and public health officials campaigned for regulation.

The universities of Cambridge and Oxford already had their bizarre privileges, both in electing their own members of parliament and having their own private prostitution laws passed in 1825.†[29] The Proctors (university policemen) were allowed to arrest streetwalkers, often along King's Parade and Trinity Street, now the tourist hub of Cambridge. Over 200 arrests a year were made in the 1840s and 1850s, and these women were interrogated, registered and medically examined. If found to be soliciting, and especially if infected, they could be locked up for weeks or months in the 'Spinning House' in St Andrew's Street – on a site now just opposite Sainsbury's Local – and treated for any infections.

The notorious Contagious Diseases Acts, first passed in 1864 and amended in 1866 and 1869, took the Cambridge example and applied it to army towns and ports. They were intended to prevent the spread of venereal disease (VD) among the armed forces – police were allowed to identify women as prostitutes and give them an internal examination every fortnight; if found infected with syphilis or gonorrhoea, the women were forcibly detained in 'lock hospitals'

---

†Barnwell was the centre of prostitution in Cambridge, coincidentally where I now live, with its hub in Wellington Street, less than half a mile from my house: in the 1860s there was a string of brothels where now there are just offices and a car park.

for three to nine months. The examination was short, sharp and potentially unhygienic: Acton wrote that 'In the course of one hour and three quarters I assisted in the thorough examination of 58 women with the speculum' – that's less than two minutes each.

The most remarkable feature of the Acts, at least from a statistical perspective, is that they were introduced in a deliberately experimental manner.[30] Fourteen of the largest stations (i.e., bases), including Woolwich, Aldershot, Cork and Curragh, were selected as 'stations which came under the Acts', while a further fourteen, including Edinburgh, Dublin and Manchester, were, in modern terminology, a 'control' group named 'the fourteen stations never under the Acts'. This shows an admirable commitment to the scientific method: unfortunately the stations do not appear to have been allocated at random, and the stations under the Acts clearly contain the major garrison towns.

Even though not required under the Acts, soldiers entering the 'intervention' stations were examined for disease whereas, as Dr Nevins told the Royal Statistical Society in 1890,

> in the stations not under the Acts there was no such examination, and the men were allowed to bring in any amount of disease they had contracted during their furlough. The first women they consorted with could go and spread it, and then the thing worked its vicious round. The question was asked in the Select Committee of the House of Commons, 'why is this not applied to all the stations?', and the answer was, that a great experiment was going on, and every precaution must be taken to prevent its failing.

Figure 57, reproduced from the original publication, shows what happened in the intervention and the control group of stations, showing that the fourteen stations under

Figure 57: **Diagram showing the fluctuation of primary venereal sores from 1860 to 1889**

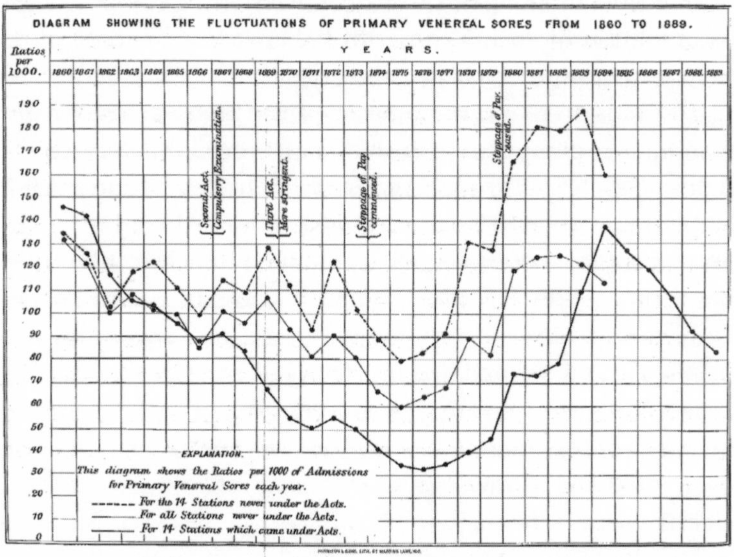

The lowest dark line is the 14 'control' stations under the Acts, the top dashed line is the 14 stations never under the Acts – the intermediate solid line is all the stations never under the Acts. Reproduced from the Journal of the Royal Statistical Society (1891).

the Acts had lower rates of primary venereal sores than stations not under the Acts. Also indicated is the order between 1873 and 1879 to stop the pay of soldiers while in hospital for treatment of venereal disease: the Minister of War was warned at the time that the only effect of this measure would be that soldiers would conceal their disease and so push rates up through lack of treatment, and this appears to have been the case.

The 'great experiment' was crucial: without it the promoters of the Acts could have claimed that the major drop in infections seen between 1864 to 1875 was due to the Acts, whereas it is clear the overall pattern was visible in the control stations as well, and all stations suffered under an epidemic from 1875 onwards: by 1885 there was little difference between the different groups of stations.

The state-regulated prostitution embodied in the Acts, which placed the blame for disease wholly with the workers rather than their clients, was deeply controversial, and there was a sustained call for their repeal. The involvement in the campaign of middle-class women such as Florence Nightingale, Harriet Martineau and Josephine Butler caused a sensation, and there were hundreds of books and pamphlets, massive public meetings and petitions with over 2.5 million signatures. And they were successful: compulsory examination of suspected prostitutes was eventually stopped in 1883, and the Contagious Diseases Acts were repealed in 1886.[†] The Acts are now mainly remembered as having given rise to the early feminist movement in Britain,[31] but it was also a revolutionary use of a controlled experimental design to evaluate a criminal justice measure, something that is now completely overlooked.

How many people have sexually transmitted infections?

25%: the proportion of 15- to 24-year-olds going for chlamydia testing each year

The importance of sexually transmitted diseases a century ago is reflected in the setting up in 1913 of the Royal Commission on Venereal Diseases. It reported in 1916 that official estimates of only around 2,000 deaths from syphilis per year were 'worthless': syphilis was rarely explicitly mentioned on

---

†Cambridge carried on locking up prostitutes until in 1891 17-year-old Daisy Hopkins sued the University for damages after her arrest for 'walking with a member of the University in a public street', supported by funds raised from the public. Although Daisy was acknowledged to be a prostitute, the case got so much publicity that the University rights were finally abolished in 1893.

Figure 58: **Syphilis diagnoses in England and Wales, 1931–2013**

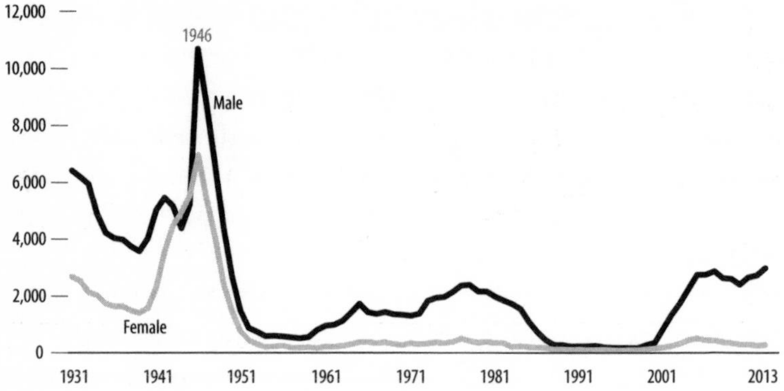

Source: Public Health England.

the death certificate through not wanting to hurt the suscep-
tibilities of relatives, and many other lethal conditions were
caused by syphilis. The only reasonably reliable statistics
were for the armed services. Based on tests in hospitals, they
concluded that 1 in 10 of the urban population had syphilis,
and this was 'greatly exceeded' by those with gonorrhoea.

By the time the Commission reported, in 1916, preparations for the Battle of the Somme were taking place and the
brothels of France had sent diseases among the troops soar-
ing. Some 23,000 British soldiers were hospitalised for treatment in 1917, which was also the year in which the first ever
sex education film was released: the Canadian silent *Whatso-
ever a Man Soweth*. It still makes powerful viewing, showing
wards of blind children born to women with syphilis. But
to put this 23,000 in perspective, this is just an average eight
days' worth of casualties in the British forces in 1917.

Sexually transmitted infections thrive in wartime. This is
dramatically shown in Figure 58, which displays the recorded
cases of syphilis back to 1931. There is a massive spike in the
Second World War, with women not lagging far behind men,
many presumably infected by servicemen. And then a steep

274

Figure 59: **Gonorrhoea diagnoses in England and Wales, 1925–2013**

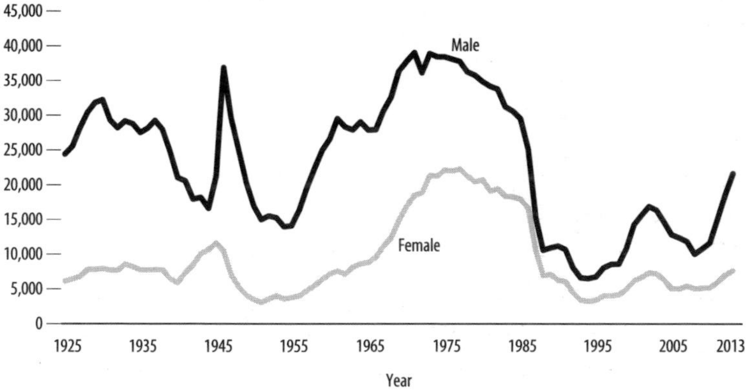

Source: Public Health England.

decline into civilian life, and an effective treatment by peni-
cillin meant fewer infectious people. There was a rise in the
libertarian 1970s – we've already seen the manifestations of
this period in teenage pregnancies – and then rates plummet
in the late 1980s as the publicity around HIV and AIDS made
safe-sex attitudes popular. There is now a resurgence, largely
driven by men who have sex with men (MSM), who make
up 74% of new diagnoses. But the numbers are still not even
approaching pre-war rates.

Gonorrhoea has occurred at around ten times the rate
of syphilis, and Figure 59 shows a broadly similar pattern
– a wartime spike, and then a peak in the 1970s that even
exceeded the wartime totals, then a plunge to the safe 1990s.
Again there is now a rise, which is partly due to increased
testing, although Public Health England suggests that there
is increased unsafe behaviour in MSM, who make up 46% of
new diagnoses.[32]

Accompanying the rise in disease has been a major change
in attendance at sexual health clinics over the last twenty
years: in 1990 Natsal-1 found that only around 1 in 30 people aged between 16 and 44 reported having been to a clinic

Figure 60: **Diagnoses of sexually transmitted infections in England in 2013**

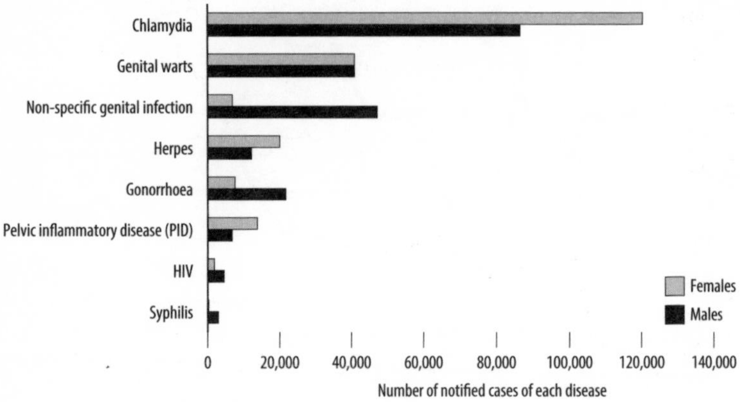

HIV numbers are for UK in 2012. Source: Public Health England.

in the previous five years. But in 2010 Natsal-3 found that around 1 in 5 (21% of women and 20% of men) had done so, and this rose to nearly half (45%) of MSM. High-risk behaviours have not increased except in MSM, and so this change is due to improved services and willingness to be tested.

Figure 60 shows the 'league table' for diagnoses of sexually transmitted infections and HIV in 2013.[†] Chlamydia comes a clear top, particularly for young women. There is a large reservoir of disease in the population, and the amount found depends crucially on how hard you look for it: in 2013 there were 1.7 million tests in the 15–24 age group – around 35% of young women and 15% of young men – an amazing example of mass screening. Of these, 139,100, around 1 in 12, were positive. When it comes to trends over time, chlamydia is increasing, genital warts have plateaued and herpes is still going up, though, again, much of the increase is explained by more people going for testing.

---

†For all except chlamydia, this only includes diagnoses made in genitourinary medicine (GUM) clinics, and so will omit any made elsewhere, such as general practice.

Figure 61: **Annual new HIV and AIDS diagnoses and deaths in the UK, 1981–2012**

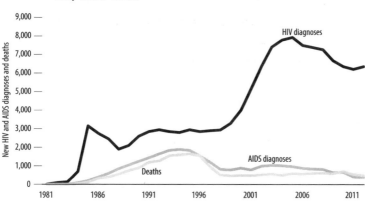

Source: Public Health England.

Although still the subject of substantial social anxiety, HIV has not fulfilled some of the dire predictions made in the AIDS 'tombstone' advertising campaign of 1987, in which John Hurt said in sepulchral tones, 'If you ignore AIDS, it could be the death of you.' At the time there was profound fear of infection, amid lurid rumours of how you could catch it from sharing a glass.[33] In 1988 a team of statisticians led by Professor David Cox predicted between 10,000 and 30,000 cases of AIDS by 1992, which was the best they could do based on limited data and understanding of the disease.[34] They said the lower projection was more plausible – and they were right; it was just slightly above the 1992 cumulative total of 9,700 cases.

Figure 61 shows the new cases each year: the feared heterosexual epidemic did not take off, and effective treatment came along; with anti-retroviral therapy (ART), people diagnosed early can expect a near-normal lifespan. There were 6,360 new diagnoses of HIV in 2012 in the UK, and 490 deaths.[35] The recent decline in HIV diagnoses is largely due to fewer reports from black African men and women, although the number of diagnoses among other heterosexuals is also falling, as a result of fewer being infected in other countries. The number of new infections among injecting drug users is fairly small – around 120 out of the 6,360 – but recorded infections among MSM are rising.

It's estimated that nearly 100,000 people are living with HIV in the UK – but nearly a quarter, around 22,000 people, don't know it yet. Around 40% of these are MSM, and around 30% black African men and women. Although Natsal-3 found only three women and six men who were HIV-positive, all either MSM or black ethnic origin, the information that the surveys provide about behaviour is exactly what is needed by the complex statistical models used to estimate this pool of hidden infections. But the models also need estimates of the chance of infection.

What's the chance of getting infected?

54: out of a crew of 1,896, the number of US sailors catching gonorrhoea during four days' shore leave in the Philippines

Those responsible for preventing teenage pregnancy are not very keen on advertising the chance of getting pregnant from a single occasion, even though the maximum 50% that we saw in Figure 46 on p. 197, based on the European Study of Daily Fecundability, does seem rather dangerous odds to play with. There's a similar anxiety about talking about the chance of getting infected with a disease from a partner: if these seem low, maybe it will encourage unsafe sex.

But another reason for not bandying these numbers around is that it's not clear what they are: even averages are difficult to estimate, let alone how the specifics of who you are, what you've got lurking inside you and what you do will affect the risks. Here goes anyway, but these are definitely not to be used for guiding your behaviour.

For vaginal sex, recent reviews put the average risks of male-to-female HIV transmission in high-income countries at 0.08% from having sex once, and female-to-male at half that risk,[36] but with a lot of variation between studies. These are mainly based on following up 'discordant' couples in which one is infected and the other not, and checking if the partner becomes infected. It's rather similar to estimating the chance of getting pregnant at a single attempt, and similarly complex. In particular, it appears the chances vary substantially between couples; of females who had unprotected sex thousands of times, the majority remained uninfected, whereas 10% of those who had had sex fewer than ten times got infected.[37] So any 'average' probability is rather misleading, and it would be misleading to report 0.08% as 1 in 1,250 occasions, as this will represent the experience of few people.

For anal sex, insertive-to-receptive transmission of HIV is estimated at 1.4% per act, while receptive-to-insertive is again lower, at 0.11% for circumcised and 0.62% for uncircumcised men. So anal sex is around twenty times as risky as vaginal sex.

Gonorrhoea is far more easily transmitted than HIV – the chance of being infected is estimated to be more than 50% if you are the receptive partner. The risk of a man being infected from a woman is lower, estimated to be around 20%, and this figure is largely based on a rather unusual, and unrepeatable, study.[38]

In the 1970s a US aircraft carrier in the Western Pacific went into Subic Bay in the Philippines for four days. When the crew returned to the ship, 54 had acquired gonorrhoea during their 'rest and recreation' in nearby Olongapo, home of around 8,000 registered 'hostesses'. These constituted 10% of the 537 men who acknowledged they had had sex with a hostess (28% of the crew of 1,896), on average four times each, and of whom only 29 had used condoms. For example, of 78 men who said they had had sex only once, 4 were infected (5%).

But to estimate the transmission risks, we need to know how many of the hostesses were infected. Rather like under the old Contagious Diseases Acts, the hostesses were required to be examined every two months, and of the 511 examined during the boat's visit, who came from 35 of the 200 local registered bars, 90 (18%) had gonorrhoea. Assuming these women were similar to those the sailors visited, a fairly simple statistical model estimated the chance of infection from a single 'exposure' of 19% for white men. There is also quite a major assumption that all the sailors were honest about their behaviour: my judgement is that the 20% figure is on the high side. With subsequent fear of HIV, it is unlikely such a 'natural experiment' would occur now.

The reason why researchers seem so obsessed with how many partners people have is that disease loves a nice diffuse network of people having sex with each other. If you're on Facebook, think of all your 'friends', and their 'friends' and so on, forming a web spreading throughout the world. And then imagine if all these people had had sex with each other, rather than just being accepted as a 'friend' on a social network. It would not take a virus or bacteria long to go through the population.

You could get an idea of how many people you are sexually connected to by putting your partner history into a (rather dubious) 'Sex degrees of separation' online calculator: for example, if you are a 35-year-old woman, who has only ever had sex with her husband of the same age, you may be surprised to know that you have 1,078,127 indirect partners six 'generations' out, and 988 three 'generations' out. Maybe time to ask your husband some questions.† These seem fairly

_____

†The program is based on a 2009 YouGov survey of 6,000 people which provided estimates of numbers of partners in different age groups. These are then multiplied out through a network of assumed 'average' people, although no account is taken of timing. The numbers quoted for the 35-year-old couple suggest an average of 10

1* stats, but did provide the *Daily Telegraph* with the fine head-line 'Brits have "indirect sex" with 2.8 million people'.

Of course, there is risky behaviour about: of 25- to 34-year-olds responding to Natsal-3, 10% of men and 8% of women reported having at least two sexual partners without using a condom in the last year. It's exactly this sort of information that is vital in working out how infections spread through a population.

However, while prostitution and disease are undoubtedly serious, the final topic of this chapter – coercion into sex – is not only the darkest issue I will cover, but perhaps also the most statistically challenging.

How many people have sex against their will?

1: out of 75 rapes in England and Wales, the number that end in a conviction

Statistics are inevitably a bare summary of a mass of personal histories, but they seem particularly inadequate when they concern people having sex against their will. Whether sexual assault, rape, child abuse and exploitation, trafficking or coercion in any way, the numbers are almost embarrassingly unable to convey the experiences people have gone through. To make matters worse, statistics in these areas are generally of poor quality, even though they are used extensively in public debates.[†]

It hardly needs saying that this is a very difficult area to research. Defining 'sex against your will' is a problem in

---

partners per 'generation' is being used, since 988 = 10 × 10 × 10, and 1,078,127 = 106.

† Although they are vital topics, I will not be covering child abuse and exploitation, violence against sex workers, child marriage, female genital mutilation or sexual violence in war.

itself: 'forced sex' suggests a degree of physical violence that may not have taken place, and words such as 'rape' may lead to lower reporting as victims may not identify their experiences with this term. Then stigmatisation of the victim, combined with unsympathetic or disbelieving authorities, can promote chronic under-reporting. Different data collection methods can give very different answers.

Starting with the official crime figures, there were 64,200 sexual offences recorded by the police in England and Wales in the year up to March 2014, which was a steep (20%) rise over the year before, while the 20,725 recorded rapes showed a 27% rise.[39] But we don't know how much of this increase is due to more events, victims being more willing to come forward after recent high-profile celebrity cases, or increased willingness of the police to record the event as a sexual offence or rape.

And this is an area where the 'official' statistics are definitely not 4*. The UK Statistics Authority, the official watchdog for government statistics, gives its own form of star rating, designating as 'National Statistics' those numbers felt to be of sufficient quality. In a dramatic move, in January 2014 this designation was removed from police-recorded crime data, after concluding that forces were not providing trustworthy counts.[40]

Instead, data from the Crime Survey for England and Wales (CSEW) are considered more reliable.[†41] This estimated that 1 in 40 women (2.5%) and 1 in 250 men (0.4%) were the victim of a sexual offence in the previous year, corresponding to 404,000 women and 72,000 men. Rather more than the police recorded. For the most serious offences of actual or attempted rape or 'sexual assault by penetration' they estimated 1 woman in 200 (0.5%) had been a victim the previous

† The 2012/13 CSEW carried out a face-to-face survey of around 50,000 households, with a self-completion module on intimate violence asked of adults aged 16 to 59.

Figure 62: **Estimated numbers of rapes each year in England and Wales between 2010 and 2012, and what subsequently happened**

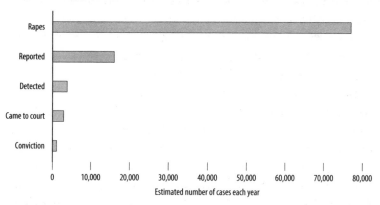

Source: Office for National Statistics (ONS).

year, coming to around 85,000 cases, around 90% of whom knew their attacker. One in 1,000 men (0.1%) had also been a victim.

The ONS estimates that each year between 2010 and 2012 there were 60,000 to 95,000 rapes, and that around 16,000 were reported to the police, of which 3,800 were 'detected' (i.e., an offender was identified and charged or cautioned), 2,910 came to court, and there were 1,070 convictions.[42] Figure 62 reveals this dramatic picture of failure, with only around 1 in 75 rapes leading to a conviction.

An alternative source to the Crime Survey for England and Wales is Natsal. The Natsal-3 team in 2010 used the term 'non-volitional sex' to make it clear they were including experiences that were not necessarily a criminal offence, asking respondents whether anyone had ever 'tried to make you have sex, against your will', and if so, had they 'Made you have sex, against your will' – this was part of the self-completion computer interview which was locked and invisible to the interviewer.[43]

Table 6: **Respondents ever experiencing sex against their will**

	Women	Men
% experiencing attempted sex against their will	19%	5%
% experiencing completed sex against their will	10%	1.4%
Median age at which it last occurred	18	16

Source Natsal-3, 2010.

The results are shown in Table 6. One in 5 women had had someone attempt to make them have sex against their will, and it was completed for 1 in 10, at a median age of 18 when it last occurred. One in 20 men had had it attempted, and one in 71 completed, at a median age of 16. We would expect the rates for men to be lower than for women, but they still form a substantial minority, and tended to have last occurred at a younger age. Much higher rates are recorded for people who were in care: 37% of women had an attempt to have sex with them against their will 'completed' and 10% of men, although the numbers are small. And rates were higher in women with poor health, early sexual experiences, more partners, low sexual function and so on.

The Crime Survey for England and Wales estimates that 19% of women and 3.4% of men are subject to sexual assault in their lives, which is close to the Natsal figures for attempted non-volitional sex, even though the definitions and age ranges differ. But the Natsal estimates for completed non-volitional sex are about double the Crime Survey rates for rape: this is probably due to the Crime Survey using the term 'rape'.

But Natsal and the Crime Survey agree on two features. First, the perpetrator is fairly rarely a stranger: only in around 15% of cases. In victims under the age of 16 it is almost always a member of the family or an acquaintance, while for older victims, above 25, it is primarily a present or past intimate partner. Second, when it comes to reporting, fewer than half the victims told anyone, and only 13% of the women and 8% of men told the police – astonishingly, even when the perpetrator was a stranger, the reporting rate only rose to 21%.

Coercion does not necessarily mean physical action or threat. Kaye Wellings feels that many teenage girls are subject to unwanted advances bordering on coercion, and 'doing things they don't want to do to please men'. Pressure can be exerted by a peer group or wider society, or even by girls themselves, internalising an external idea of how they should be behaving. All of which can have serious consequences for the girls' mental state and future.

Can we compare statistics on rape?

66: the recorded number of rapes each year per 100,000 people in Sweden
2: the recorded number of rapes each year per 100,000 people in India

If British sexual coercion statistics seem poor, trying to make international comparison is a statistical minefield. Take rape statistics. The country that has by far the highest reported rate is Sweden, with 66 victims per 100,000 inhabitants per year in 2012, compared with the UK official rate of 24 per 100,000 a year. But is Sweden so much more dangerous than the UK? I think not. Sweden's reporting system is specifically designed to bring attention to these offences: multiple events with the same victim are recorded as separate crimes, there is a wide definition of rape and reporting is encouraged. In contrast, India, which has a notorious reputation for sexual violence against women, has a tiny rate of 2 per 100,000 people a year.

We've seen in Chapter 11 how the World Health Organisation (WHO) has made clear that sexual health is more than just a lack of disease: it concerns our whole well-being. The

WHO says that 'Sexual health requires a positive and respectful approach to sexuality and sexual relationships, as well as
the possibility of having pleasurable and safe sexual experiences, free of coercion, discrimination, and violence.'[44] These
are fine principles, but a large international review estimated
that one in fourteen women (7%) in the world had been the
victim of sexual violence from a non-partner, although not
necessarily a stranger,[45] and rates rose to around one in five
in sub-Saharan Africa. If we take the WHO sentiments seriously, then this means changing what we mean by promoting
sexual health, and not just using rates of teenage pregnancy
and disease: statistics on coercion, and even satisfaction, are
necessary.

Young people are seriously concerned about coercion. In
the recent Institute for Public Policy Research survey of 500
18-year-olds, they were asked 'What do young people need
to know about, discuss or understand to have healthy, happy
and positive relationships?'[46] The top response, mentioned
by 75%, was 'Reproduction, pregnancy, safe sex, sexual diseases and sexual health checks' – the standard sex education
material. But next on the list, with 68%, was 'Getting and
giving consent'.

Just as with pornography, maybe we should just listen to
young people a bit more.

15

A BOY OR A GIRL

Are more boys born than girls?

~~~~~~

21: the number of boys born in England
and Wales for every 20 girls

~~~~~~

Couples, whether peasants or royalty, have always observed the swelling bump and hoped, for whatever reason, for a boy or a girl. Historically, the wish would often have been for a boy, especially a king wanting an heir, and this explains why there is plenty of advice out there about how to increase the chances of a boy. You could try having sex with the woman on top or doggy-style, or eating a lot, especially bananas. But if you'd prefer a girl, what about having sex some time before ovulation, as the strong, slow female sperm will outlive those feeble male efforts? But articles generally conclude that the chance of having a boy or a girl is just like flipping a coin: it's 50:50. But is it?

In 2012 in England and Wales, 374,346 boys were born but only 355,328 girls – an excess of 19,018 boys.[1] We could also say that 51.3% of live births are boys, which sounds like rather a biased coin. Another way of expressing this imbalance is through the 'sex ratio' at birth, generally defined as the number of males born per 100 females. So for England and Wales in 2012 this is 105.4[†] – it generally moves around 105, so there are 5 extra boys born for every 100 girls, or equivalently, 1 extra boy for every 20 girls. This is known as the 'secondary' sex ratio as it reflects the imbalance in registered live births. The 'primary' sex ratio is that at conception and is even higher, as male foetuses are more fragile than female ones.[2]

And the male sex continues to be weaker after a baby is born: each year an average male, of whatever age, has around a 50% higher chance of dying before his next birthday than a woman of the same age. This extra 'force of mortality', as the annual fatal risk is rather archaically known, translates to males, on average, having the same annual fatal risk as women four years their senior, and so men live on average four years less than women. So much for the weaker sex.

It can be tempting to read some sort of mystical intervention into the excess of males at birth. And this is exactly what early investigators did. We've seen that parish registers were introduced in 1538, and they were followed in 1604 by weekly and annual 'Bills of Mortality' for London, soon after James I became the first king of both England and Scotland (and Wales). These also included christenings, and after the sex of the christened baby was added in 1629, one of the first exercises in Big Data could get under way.

The first analysis took place in 1662 by John Graunt, a hero among statisticians, who produced huge and wonderful tables showing for each year how many people died of various causes: for example, in 1660 there were 31 deaths put down to the French pox (syphilis), 54 from the King's Evil (scrofula) and 402 from 'plague in the guts'.[3] But he also noted that between the years 1629 and 1661 there had been 139,782 males and 130,866 females christened in London,

---

† $100 \times 374{,}346 / 355{,}328 = 105.4$

giving an overall sex ratio of 107, not far from the current figure.

Graunt estimated that for every 13 females, 14 males were born, and concluded that this extra 1 in 13 was needed to prevent polygamy:

> more men die violent deaths then women, that is, more are slain in Wars, killed by mischance, drowned at Sea, and die by the Hand of Justice. Moreover, more men go to Colonies, and travel into foreign parts, then women. And lastly, more remain unmarried, then of women, as Fellows of Colleges, and Apprentises, above eighteen, &c. yet the said thirteenth part difference bringeth the business but to such a pass, that every woman may have an Husband, without the allowance of Polygamy.

If Graunt is a hero to statisticians, John Arbuthnot is a super-hero. He was a modest but accomplished man: apart from being made physician to Queen Anne in 1705, he was also an inspiring friend to Jonathan Swift, John Gay and Alexander Pope, and as part of this satirical circle invented the figure of John Bull. He had the robust sense of humour characteristic of the age. When he translated Christiaan Huygens's *Laws of Chance* from Latin in 1692 (producing the first book on probability published in English), his additional gambling examples in the introduction referred to the high rates of promiscuity and the pox: 'It is hardly 1 to 10, that a woman of Twenty Years old has her maidenhead, and almost the same wager, that a Town-Spark of that Age has not been clap'd.'

Sadly he destroyed all his personal writings, and so his main legacy is a short paper published in 1710. But what a paper: it is now seen as the very first example of statistical inference, the fundamental and vital idea of using the theory of probability to draw conclusions about hypotheses on the basis of observed data.[4]

His data concerned the sex ratio. He extended Graunt's analysis to 1710, and showed that in each of 82 years there had been more males than females born, with an overall sex ratio of 107, varying between 101 and 116 over the period. He then argues that, if the true chance of a male were really 50:50, there is only a tiny probability that every single one of the 82 years would show such an excess in this one direction.[†] He therefore concludes that some force is at work to counter the excess mortality of males: 'To repair that Loss, provident Nature, by the Disposal of its wise Creator, brings forth more Males than Females; and that in almost a constant proportion.'

The title of his famous paper uses the data as direct statistical evidence for the existence of supernatural intervention: 'An argument for Divine Providence, taken from the constant regularity observ'd in the births of both sexes.'[5]

When did baby boys most outnumber baby girls?

106.5: the highest sex ratio recorded since 1838 in England and Wales, in 1944

Whether we believe in Divine Providence or not, we do have to acknowledge that something causes a consistent excess of males, but the ratio is not constant. Figure 63 shows the sex ratio in England and Wales from 1838 to the present day: I have never seen this plotted before, and it is one of the most extraordinary graphs I can think of.[6] These are not just random fluctuations: for some reason the ratio of boys to girls has systematically changed over the last 200 years. And the pattern has some very curious features.

---

† The probability is $1/2^{82}$, which is a very small number with 24 zeros after the decimal place. Arbuthnot's data have been subject to repeated analysis, and although there may be counting errors, and only Anglican baptisms are included, the basic finding still holds.

Figure 63: **The number of boys born for every 100 girls in England and Wales, 1838–2012**

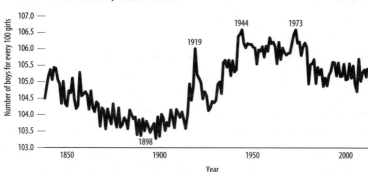

Source: Office for National Statistics.

First we note a steady decline in sex ratio throughout the Victorian period, reaching a minimum at around 1900, and we shall come up with a rather speculative explanation for this in due course. The sex ratio then increases, reaching a plateau between 1950 and 1970, and then declines again to the current levels after around 1980. But among these general trends there is some remarkable behaviour.

Crucially, there are some dramatic peaks: during and just after the First World War (1914–18), with a sharp rise to 106 in 1919, and also a peak around and just after the Second World War (1939–45), with the sex ratio in 1944 (106.5) being the highest since modern records began. We'll come to the 1973 peak later.

What is going on? Why should more boys be born at the end of wars? Is some mystical force replacing the males lost in warfare with new baby boys, ready to do their bit when the time comes again? There's been extensive research around this strange phenomenon, with a sex ratio peak in 1919–20 being claimed for Germany, Austria, Belgium, Bulgaria, France, the UK, Hungary, Italy, Romania and South Africa – all nations that had been fighting in the war.[7] The USA also saw its highest sex ratios during and just after the

Second World War, with 1946 hitting 105.9.[8] As we'll see, a variety of (non-supernatural) explanations for this mystifying phenomenon have been suggested.

Why are more boys born at the end of wars?

56%: the proportion of girls born to parents rated 'attractive' as adolescents (1*)

Other European countries show sex ratios of between 105 and 107,[9] and so it's a robust finding that on average the odds are slightly in favour of having a son. But what can make the sex ratio higher or lower than 106?

Almost everything you can think of has been found to have some relation to the sex ratio: for example, race (e.g., high for Chinese, low for black), mother's and father's age (younger have more boys), birth order (more boys early on in the family), war (more boys, as we saw above), handedness (more boys for right-handed parents), 'coital rate' (more sex leads to more boys), exposure to chemicals and so on and so on.[10]

Increased parental stress has been associated with having more girls, such as following the 1995 Japanese earthquake or the collapsing East Germany economy in the 1960s: the question arises whether this might be because fewer boys are conceived, or more than usual die while they are still a foetus. The destruction of the Twin Towers on 11 September 2001 provided an opportunity to study this by examining the sex ratio of births in New York in the subsequent months. This fell to 100 in January 2002, three months after the attack, whereas no change was seen nine to eleven months afterwards.[11]

This suggests that stress during pregnancy may lead to more miscarriages of male foetuses than female. The same researchers looked directly at the sex ratio among foetal deaths in California and found that relatively more male foetuses died when the unemployment rate was higher. They estimated that 370 male foetal deaths in California could be attributed to the economic downturn between 1989 and 2001, although there must be a lot of uncertainty around this estimate.[12]

Many of these associations are small and questionable, with inconsistent evidence. Some claimed effects are large and even more questionable: a classic being the idea that beautiful people tend to have more girls. This is used as evidence for the 'Trivers–Willard hypothesis': the theory that evolution has led to mammals adjusting their sex ratio to the gender most favoured by the parents' condition – so, for example, high-ranking monkeys may produce more males than those who are low-ranking. So it's been claimed that violent men have more sons,[13] engineers have more sons and nurses have more daughters.[14] Or beautiful people should have girls, because they will be beautiful too and succeed in life.[15] If you're ugly, you'd be better off having a boy, as his appearance will not have so much effect on his prospects. Convinced? No, neither am I.

But at first sight the evidence about attractive people having more daughters looks quite strong. One study looked at participants in a US survey of adolescents who at their initial interview had their 'physical attractiveness' rated on a five-point scale (it must have seemed a good idea at the time).[†] Fifteen years later nearly 3,000 had gone on to have a child, and of the first-born of those who had been rated as 'very attractive' as an adolescent, only 44% were boys, as opposed to the standard 52% for all the plainer people.

[†] As Francis Galton found out when he was wandering around Britain, it's difficult to measure attractiveness. It's been suggested that the correct unit is the 'milli-Helen', which is the amount of beauty needed to launch one ship.

This is apparently a very powerful result and got a lot of media attention, particularly as the abstract of the paper and subsequent coverage wrongly claimed that very attractive individuals were '26% less likely to have a son'.[†] But the statistician and blogger Andrew Gelman of Columbia University has pointed out some curious features of this analysis.[16] If we look at the proportion of boys in all five categories individually, we find the first-born children were:

- for parents who had been rated as 'very unattractive' as adolescents: 50% boys
- for those rated 'unattractive': 56% boys
- for those rated 'about average': 50% boys
- for those rated 'attractive': 53% boys
- for those rated 'very attractive': 44% boys.

So there was no steady decline – only the 'very attractive' category showed any sign of a reduction. Gelman argues that not only does this suggest that the results have been selectively reported to tell the right 'story', but that the 44%, which corresponds to a sex ratio of 79, is in any case implausibly extreme.

But does this help us explain the extra boys born during and after wartime? If we believe Trivers–Willard, we might be able to argue that evolution has enabled women to have more boys as they will do better in a society short of males – we don't need to bring God into it, just Darwin. But what about the extra boys born to younger parents, early on in marriage, and to those who conceive quickly?

---

† It seems that the '26% reduction' in fact refers to the 'odds' of having a boy, which were reduced from 52/48 to 44/56. So the data could be interpreted as an apparent 26% reduction in the sex ratio.

Why does more sex mean more boys?

1898: the year when British sexual activity
may have been at its lowest ebb

To me the most coherent explanation that brings many of these findings together is that of Professor William James of University College London. He has been arguing for around forty years that the sex of the foetus is influenced by the hormone levels of the parents at around the time of conception, and in particular that the sex ratio varies over the monthly cycle, with conceptions early, and possibly also late, on in the cycle having a greater tendency to be males, and more females arising from conceptions around ovulation.[17]

James claims strong indirect support for this idea from evidence that increased 'coital rate' – that is, more intense sexual activity – leads to slightly more boys being conceived and born to younger couples. But why should having more sex lead to more boys? We saw in Chapter 10 that the peak fecundability was around two days before ovulation, but that if couples had a lot of sex they were more likely to conceive before this peak, as the chances were higher that the woman would already be pregnant by the time her peak fertility arrived. In particular, during and just after major wars, sex has to be crammed into brief periods of leave, whatever the time of the month, or is intensely experienced in the exhilaration of being reunited following demobilisation. This makes it more likely that any resulting conception will occur earlier in the fertile period of the cycle, and so, according to James, be more likely to be a boy.[†]

Direct evidence is harder to come by, as the day of

[†] A simple statistical model by James and Valentine estimates that babies born after a three-day leave, with sex on every day, could on average have a sex ratio of around 112.5.[18]

conception relative to ovulation has to be known. And where could we find thousands of women who would know almost exactly when they had conceived? One answer came to Jerusalem researcher Susan Harlap, who realised the tradition of *niddah* by orthodox Jewish women provided a unique opportunity.[19] This practice involves a taboo on sex for at least a week after the end of menstruation, following Leviticus 15:19: 'And if a woman have an issue, and her issue in her flesh be blood, she shall be put apart seven days.' After seven days the woman has a ritual cleansing bath (*mikve*), and then sex can start again.

Harlap studied 3,658 births to women following this tradition, estimating from their reports the day of conception relative to ovulation. She found that of babies conceived before ovulation 53% were male, of those conceived on or the day after ovulation 50% were male, and of those conceived two days after ovulation 65% were male. This pattern held for women of different ages and ethnic backgrounds. This was not a perfect study since the day of conception could only be inferred and will have some error attached to it, but she argued that this meant the true relationship must be even stronger.

If we take this idea seriously, then the opposite will hold: if there is less sex going on, then this should mean conception tends to occur around the most fertile time of the month, and so there would be more girls and the sex ratio dropping. And so now, finally, we come back to the mysterious fall in fertility of the Victorians that we saw in Chapter 9, where they stopped having so many babies in spite of not using contraception, and how historians such as Simon Szreter have argued that this happened because they developed a 'culture of abstinence' – that is, they just didn't have as much sex. If we believe the James hypothesis, or even just accept that sex ratios are in some way influenced by 'coital rate', then it should be clear that the graph of sex ratios over time can be taken as a proxy measure for the intensity of sex going on.

Look again at Figure 63, and what happens over the Victorian era: the steadily dropping sex ratio might provide additional statistical support, based on 4* data, to the 'abstinence' idea – there was simply a dearth of sex. We could therefore estimate that the historical nadir of British sexual activity occurred around 1898. And then peaks of sexual intensity occurred at the end of wars with the returning troops, and again around 1973, at a time when we've seen that the marriage age reached its lowest point and there was a surge of teenage pregnancy – it was clearly a time of intense sexual activity in the young. Frantic fornication breeds boys. My case rests.[†]

Why do some countries have more boys?

118: the number of boys born to every 100 girls in China

International data suggest variation between countries and over time but that, in the natural course of events, long-term sex ratios at birth outside the range 103 to 108 would be unusual. But events are not necessarily natural, and if there is a strong gender preference, it may be possible to have a blood test or an ultrasound, find the sex of the foetus and then arrange an abortion if it is not what is desired (which usually means it is a girl). This practice, although illegal, led to sex ratios at birth of 114 in South Korea in 1992, rising to around 200 for third and fourth births – an incredibly high figure that means that, once a family had two children, half of all female foetuses were being aborted.[20] These extreme ratios have reduced following government campaigns and suspending licences of doctors carrying out sex determination, but high

---

[†] Although in fairness I should add that Andrew Gelman thinks that this is more plausibly the effect of healthy young mothers.

rates of between 114 and 126 were reported for some Indian states (Punjab, Delhi and Gujarat) in 2001.

China's One Child Policy and strong preference for sons has led to sex ratios up to 130 in some rural areas, and 1 million excess male births are reported each year – or equivalently, 1 million missing girls. The policy ended in 2013, allowing two children if either parent is an only child, but it remains to be seen whether this will improve China's sex ratio of 118 in 2011, the highest in the world.

The UK Department of Health recently carried out a rather delicate investigation: they looked at whether sex ratios of children born in the UK depended on the country the parents were born in. The overall average, based on nearly 4 million births between 2007 and 2011, was 105: Australia, China and the Philippines had ratios above 108, but when allowing for the fact that there were 160 countries, we would have expected this by chance alone: high sex ratios have also been reported in Chinese and Filipino immigrants to the USA. At the other extreme, the sex ratio was below 103 for seven countries, but only Sri Lanka stood out as being surprisingly low, with a sex ratio of 99.2. And this is in the direction of *fewer* boys, and so suspicions were not raised.[21] But the crucial indicator is the sex of children born subsequent to having a girl: if families really want a boy, then they may resort to selective abortion at that stage.

Can you make sure you have a boy?

$20,000: rough current cost of gender selection

So, you've had a little girl, and you would like to round off your perfect family with a boy. Or vice versa. What can you do to increase your chances of getting what you want? Let's start at the highest-tech and work downwards.

During in-vitro fertilisation (IVF) the eggs can be sorted into the right sex before implanting – this is known as pre-implantation genetic diagnosis. As its name suggests, it was developed to allow couples with a genetic disease to screen their eggs, and in the UK this is its only legal use. But if you have around US $20,000 you should be able to find a US gender-selection clinic willing to do this for you, and a child of the right sex is just about guaranteed.

Less reliable methods try to separate male and female sperm before either IVF or intra-uterine insemination: the MicroSort method claims a 90% success rate. It is still not approved by the Food and Drug Administration but is offered in various clinics around the world, reportedly for about $20,000 a cycle. The Ericsson method is cheaper, around $600, and claims 80% success, but this is disputed.

Numerous websites will take your money and send you advice and kit to help time conception relative to ovulation: the Shettles method is popular and recommends having sex as close to ovulation as possible if you want a boy, based on the idea that the lighter male sperm swim faster and hence get to the egg first. Since this is the exact opposite of the evidence I've just presented, it is perhaps unsurprising that the Shettles method does not seem to work. James reports three studies of this method that, out of 131 attempts to control the sex of the child, gave rise to 57 babies of the desired sex, and 74 of the wrong one; this does not seem a great success rate.[22]

So if you feel like having a boy, and you believe the arguments of William James, pretend there's a war on and have a lot of sex early on in the cycle, when you are less fertile: it may take longer to conceive, but you may (very slightly) increase your chances of a boy. Perhaps it's better than eating lots of bananas, but it's still hardly more predictive than flipping a coin.

16
CONCLUSIONS

I tend to read biography and history. This may have been clear from the content of this book, since those perspectives are so clearly relevant to the process of trying to look into people's private sexual lives, both now and in the past. When going through the statistics, it is easy to slip into treating the data as referring to some activity that is 'out there', like observing ants or planets (or gall wasps), rather than those soft and extraordinary human things.

There are three human elements behind the stats. First, the researchers. There's Dr Clelia Mosher collecting her intimate and moving case histories, then hiding them in her papers and looking after her mother, Magnus Hirschfeld not daring to return to his Institute in Berlin, Alfred Kinsey obsessively trying to complete his work in spite of being desperately ill. Even Shere Hite, in spite of her 1* statistics, deserves great respect for her work in producing such influential books.

Then there was the forced marriage of the medical work of Anne Johnson and the social-science approach of Kaye Wellings to produce Natsal. This has been a fine and lasting achievement, necessarily starting from a strict 'health' agenda, but now steadily expanding into the broader definition of sexual health promoted by the WHO.

The second group that deserves thanks are the study participants who have contributed the most intimate details of their lives to research about sex: the total mentioned in this book must run into hundreds of thousands. They have been open about behaviour that they may never have admitted to anyone else, and if they sometimes find it difficult to count up their partners, they deserve our sympathy. And let's not forget those who volunteered to be wired up to (apparently real) lie-detectors and to recall their sexual experiences, or to watch erotic films and then plunge their hands into a bowl of (apparently used) condoms and so on. They may just have been psychology students earning research credits, but they deserve our gratitude.

But the third, and far more important, human element is the vast number of people that, for countless centuries, have had to deal with their sexuality. It's a great leveller – whether peasants or royalty, they have had to negotiate their desires and difficulties with both themselves and others. It's not for me to say whether men or women have had a more or less difficult time of it, although, apart from the last half-century, fallible contraception has meant that the risk of pregnancy will have loomed large in the mind of the woman. And this does not even begin to touch on what those who wished to, or furtively did, engage in same-sex behaviour had to endure.

The Natsal surveys have shown that we have become more sexually adventurous, while also claiming increasing tolerance of same-sex activity and intolerance of infidelity. But we should not simply see historical trends as an inevitable progress away from repression and towards enlightenment. Many people in the past will have got on quite happily with expressing their sexuality, while today, although the statistics show new opportunities for experimentation are being taken up with relish as part of the continued move away from sex being seen as 'procreative', there are also new pressures coming from pornography, coercion and the premature sexualisation of young people.

We've repeatedly come up against the competing views of researchers about whether it is 'nature *or* nurture' that determines how you feel and what you do. If you believe that sexuality is an innate biological characteristic, then the massive changes in sexual behaviour over the past half-century or so could be seen as a liberation of behaviour that has previously been suppressed. Alternatively, you may believe that a new sexuality is being created by prevailing circumstances. Or you may even think it's a complete waste of time to talk about it.

My personal view is that our sexual feelings and behaviour come from an irretrievably complex mixture of biology, society, opportunity and pure chance. And that's what makes it all so fascinating.

NATSAL METHODS

23%: the proportion of subjects who were 'embarrassed' answering questions in Natsal-1

Even if a good random sample is targeted, they still need to be persuaded to bother to respond. And response rates for all surveys are going down, even if they do not contain sexual questions. Natsal have found that a good introductory letter is vital, stressing the benefits to the health of society from the data: they also experimented with giving £30 rather than the standard £15, but this only made a marginal difference.

Natsal-3 had some deficit of single people and Asians.[1] But the crucial thing is whether the people who volunteer to take part have a systematically different sex life from those of a similar age, gender and ethnicity who refuse to participate. This is known in the statistical trade as 'informative missing data' – does the fact that someone refuses to tell us anything tell us something about what they would have said, had they been persuaded? If you see what I mean.

But how can we work out whether the sexual activities of those who refuse are different when, by definition, we don't have that information? The natural approach is to go back to a sample of the missing people and make a super-strenuous effort to find out something about them. The US National AIDS Behavioral Survey did just this: in 1990 they tried telephone contact with 13,690 randomly selected numbers, and obtained a 66% response rate – about normal for this kind of survey.[2] They then tried to find out more about those who were contacted and refused, and those who could not be contacted even after up to seventeen attempts: amazingly, they managed a 33% response rate to this follow-up study. Those who refused tend to be more conservative, to go to church more often and not to trust confidentiality, and could be assumed to have lower-risk behaviour, whereas the ones who are difficult to get hold of spend less time at home and have more sexual partners, and so are higher-risk.

Other studies reinforce the impression that those who refuse to participate tend to be less sexually liberal, while those who are hard to contact may have higher-risk behaviours, so to a certain, but unquantifiable, extent these effects cancel each other out.[3]

It's easier to study the characteristics of people who get so far but don't complete all the questions. In Natsal-1, around 1 in 25 volunteers for the face-to-face interview refused to fill in the booklet with the more sensitive questions: these refusers tended to be older, have comprehension problems, have fewer partners or come from an ethnic minority: for example, 25% of Asian responders refused to fill up the booklet.[4] The interviewer also assessed the 'embarrassment' of the volunteer: those with no partners, Asians and the unskilled tended to be more embarrassed and also tended to have lower-risk behaviours. If people are refusing to participate because they are embarrassed, it again suggests that risky behaviour may be being over-estimated in the survey.

Maybe men and women respond differently. It seems plausible that men with a lot of experience may feel happy recounting their adventures, while those with nothing much to report may be more reluctant to 'admit' their perceived deficit, while women may respond in the opposite way, with the more active refusing to participate. This is the 'social acceptability bias', which is particularly evident when we come to the tricky question of how many partners people have had in their lifetime.

20%: the increase in the proportion of people reporting a same-sex experience when filling in a form rather than being asked face-to-face

---

It's no good collecting all these data if we can't believe what people say, so there's been a huge effort to understand how to get reliable answers in surveys: how to word the questions, the effect of using computers as well as an interviewer, or even whether to dispense with the personal interview entirely and do everything online.[5]

Let's start with wording. If you were being interviewed, would you prefer if vernacular terms were used (e.g., 'wank') or formal language (e.g., 'masturbation')? It's no good being technical if it causes confusion: Kaye Wellings reported that 'I remember a man – he was 35, he had children; and when asked what "vaginal" meant, he said, "Ooh, no, no, no. I wouldn't do that, I don't like the sound of that." It was really because it sounded so strange to him.'[6]

Kinsey embraced the vernacular, but his whole approach was to establish a personal relationship with the respondent. But the Natsal team have found that formal terms are preferable, with explanation if necessary: in focus groups of young people 'They generally agreed that "sex" would be acceptable, but words such as "shag" were not.'[7] The idea is that the interviewer should be a 'professional stranger' – unflappable and trained to deal with whatever comes up.

Even the ordering of questions can be important: Kinsey was happy to dot around his form, but modern surveys work through a strict, computer-driven schedule. Behaviour is covered before asking about attitudes; otherwise subjects might adapt their responses to fit the opinions they have given earlier.

Natsal-1, in 1990, was all carried out on paper forms, with a face-to-face interview and a booklet to fill up. People may not feel happy talking about sensitive issues even to trained interviewers: around 20% more people said they had had a same-sex experience when filling up the private booklet than when directly asked face-to-face.[8] By Natsal-2, in 2000, the booklet was replaced by a computer on to which the participant entered the more sensitive information, such as experience of heterosexual and homosexual practices, numbers of partners, paying for sex, non-volitional sex and so on: this could then be locked and was invisible to the interviewer. And by Natsal-3, in 2010, the face-to-face interview was also entered directly on to a computer.

But how can we know the effect of using a computer rather than a paper form? If the computer were a new medical treatment, a fair test of its effect would be by a 'randomised controlled trial': patients would be allocated at random to the new and the standard treatment, and any differences in outcome between the two groups could be put down to the intervention, provided enough patients were used for the play of chance to be ruled out.

The same idea of randomised trials has been used extensively in sex research. For example, during the 1995 US National Survey of Adolescent Males, nearly 2,000 subjects were randomly allocated to use either a computer or a paper form for their answers.[†9] Adolescent males reported more sensitive behaviours to the computer: 2.5% said they had ever had sex with a prostitute, compared with 0.7% on paper,

---

†The computer mode also used audio to ask the questions: subjects were randomised in a four-to-one ratio to computer (n=1361) or paper (n=368).

and the proportions who reported male–male sex was 5.5% on the computer vs. 1.5% on paper.

Many other studies have shown that people tend to admit to more 'sensitive' sexual behaviours when using a computer than when filling in a paper form. However, when the Natsal team compared computer to paper forms in 829 households when preparing for Natsal-2, they found no effect on responses, although there was more consistency and fewer missed questions.[10]

It would be far cheaper and easier to conduct the entire survey over the internet, but online panels can give unreliable results. Anne Johnson feels that, on top of the issues of who participates in a web panel, the importance and value of the survey are not communicated.

It will be difficult to replace the careful personal approach, but cost constraints may dictate changes in the future.

NOTES

1: Putting Sex into Numbers

1. Sanders, S. A., and Reinisch, J. M. 'Would you say you "had sex" if …?' *Journal of the American Medical Association*, 281 (3), 20 January 1999: 275–7.
2. Hite, S. *The Hite Report: On Female Sexuality*. Pandora, 1976.
3. Hite, S. *The Hite Report on Male Sexuality*. Ballantine Books, 1982.
4. Hite, S. *The Hite Report: Women and Love: A Cultural Revolution in Progress*. Knopf, 1987.
5. Streitfeld, D. Shere Hite and the Trouble with Numbers. Available from: http://davidstreitfeld.com/archive/controversies/hite01.html [accessed 8 September 2014].
6. Larkin P. 'Annus Mirabilis'. Available from: http://allpoetry.com/Annus-Mirabilis [accessed 4 September 2014].
7. Szreter, S., Fisher, K. *Sex before the Sexual Revolution: Intimate Life in England, 1918–1963*. Cambridge University Press, 2010.
8. Smith, R. The firing of Brother George. *British Medical Journal*, 318 (7178), 23 January 1999: 210.

2: Counting Sexual Activity

1. Mercer, C. H., Tanton, C., Prah, P., Erens, B., Sonnenberg, P., Clifton, S., et al. Changes in sexual attitudes and lifestyles in Britain through the life course and over time: Findings from the National Surveys of Sexual Attitudes and Lifestyles (Natsal). *Lancet*, 382 (9907), 30 November 2013: 1781–94.
2. Field, N., Mercer, C. H., Sonnenberg, P., Tanton, C., Clifton, S., Mitchell, K. R., et al. Associations between health and sexual

lifestyles in Britain: Findings from the third National Survey of Sexual Attitudes and Lifestyles (Natsal-3). *Lancet,* 382 (9907), 30 November 2013: 1830–44.

3. Mercer, C. Let's talk about (real) sex. TedX talk. Available from: http://tedxtalks.ted.com/video/Lets-talk-about-real-sex-Dr-Cat [accessed 12 October 2014].

4. Overy, C., Reynolds, L. A., Tansey, E. M. *History of the National Survey of Sexual Attitudes and Lifestyles: The Transcript of a Witness Seminar Held by the Wellcome Trust Centre for the History of Medicine at UCL, London, on 14 December 2009.* Queen Mary, University of London, 2011.

5. Wellings, K., Field, J., Wadsworth, J., Johnson, A. M., Anderson, R. M., Bradshaw, S. Sexual lifestyles under scrutiny. *Nature,* 348 (6299), 1990: 276–8.

6. Overy, Reynolds and Tansey. *History of the National Survey of Sexual Attitudes and Lifestyles.*

7. Wadsworth, J., Field, J., Johnson, A. M., Bradshaw, S., Wellings, K. Methodology of the National Survey of Sexual Attitudes and Lifestyles. *Journal of the Royal Statistical Society,* series A, 156 (3), 1993: 07.

8. Everitt, B. S. *History of Surveys of Sexual Behavior: Encyclopedia of Statistics in Behavioral Science.* John Wiley & Sons, 2005, pp. 878–87. Available from: http://onlinelibrary.wiley.com/doi/10.1002/0470013192.bsa284/abstract [accessed 12 October 2014].

9. Fenton, K. A., Johnson, A. M., McManus, S., Erens, B. Measuring sexual behaviour: Methodological challenges in survey research. *Sexually Transmitted Infections,* 77 (2), 1 April 2001: 84–92.

10. Davis, K. *Factors in the Sex Life of Twenty-Two Hundred Women.* Harper and Brothers, 1929.

11. Kinsey, A. C., Pomeroy, W. B., Martin, C. E. *Sexual Behavior in the Human Male.* Indiana University Press, 1948.

12. Time Out. *London Sex Survey 2013.* Available from: http://www.timeout.com/london/sex-and-dating/sex-survey-2013-how-much-sex-is-everyone-having-anyway [accessed 3 September 2014].

13. Trojan U.S. SEX CENSUS Finds Sexual Diversity and Satisfaction on Rise. Available from: http://www.trojancondoms.com/ArticleDetails.aspx?ArticleId=25 [accessed 13 October 2014].

14. Erens, B., Burkill, S., Copas, A., Couper, M., Conrad, F. How well do volunteer web panel surveys measure sensitive behaviours in the general population, and can they be improved? A comparison with the third British National Survey of Sexual Attitudes and Lifestyles (Natsal-3). *Lancet,* 382 (9907), 30 November 2013: S34.

15. Wellings, K., Jones, K. G., Mercer, C. H., Tanton, C., Clifton, S., Datta, J., et al. The prevalence of unplanned pregnancy and associated factors in Britain: Findings from the third National Survey of Sexual Attitudes and Lifestyles (Natsal-3). *Lancet*, 382 (9907), 30 November 2013: 1807–16.

3: Spin Your Partner

1. Mercer, C. H., Tanton, C., Prah, P., Erens, B., Sonnenberg, P., Clifton, S., et al. Changes in sexual attitudes and lifestyles in Britain through the life course and over time: Findings from the National Surveys of Sexual Attitudes and Lifestyles (Natsal). *Lancet*, 382 (9907), 30 November 2013: 1781–94.
2. Robinson, C., Nardone, A., Mercer, C., Johnson, A. M. 'Sexual health', Chapter 6 of Health Survey for England 2010. Available from: http:// www.hscic.gov.uk/catalogue/PUB03023/heal-surv-eng-2010-resp-heal-ch6-sex.pdf [accessed 18 September 2014].
3. Wadsworth, J., Johnson, A. M., Wellings, K., Field, J. What's in a mean? An examination of the inconsistency between men and women in reporting sexual partnerships. *Journal of the Royal Statistical Society*, series A, 159 (1), 1996: 111.
4. Brewer, D. D., Potterat, J. J., Garrett, S. B., Muth, S. Q., Roberts, J. M., Kasprzyk, D., et al. Prostitution and the sex discrepancy in reported number of sexual partners. *Proceedings of the National Academy of Sciences*, 97 (22), 24 October 2000: 12385–8.
5. Wadsworth et al. What's in a mean?
6. Conrad, F. G., Brown, N. R., Cashman, E. R. Strategies for estimating behavioural frequency in survey interviews. *Memory*, 6 (4), 1 July 1998: 339–66. Brown, N. R., Sinclair, R. C. Estimating number of lifetime sexual partners: Men and women do it differently. *Journal of Sex Research*, 36 (3), 1 August 1999: 292–7.
7. Boniface, S., Shelton, N. How is alcohol consumption affected if we account for under-reporting? A hypothetical scenario. *European Journal of Public Health*, 26 February 2013: 1076–81.
8. Baldwin, M. W., Holmes, J. G. Salient private audiences and awareness of the self. *Journal of Personal and Social Psychology*, 52 (6), 1987: 1087–98.
9. Alexander, M. G., Fisher, T. D. Truth and consequences: Using the bogus pipeline to examine sex differences in self-reported sexuality. *Journal of Sex Research*, 40 (1), 1 February 2003: 27–35.
10. Wrigley, E. A., Schofield, R. S. *The Population History of England 1541–1871*. Cambridge University Press, 1989.

11. Johnson, A. M., Mercer, C. H., Erens, B., CoS., Wellings, K., et al. Sexual behaviour in Britain: Partnerships, practices, and HIV risk behaviours. *Lancet*, 358 (9296), 1 December 2001: 1835–42.

12. Blow, A. J., Hartnett. K. Infidelity in committed relationships, ii: A substantive review. *Journal of Marital and Family Therapy*, 31 (2), 1 April 2005: 217–33.

13. Hite, S. *The Hite Report: Women and Love: A Cultural Revolution in Progress*. Knopf, 1987.

14. Wolfe, L. *The Cosmo Report*. Arbor House, 1981.

15. Wiederman, M. W. Extramarital sex: Prevalence and correlates in a national survey. *Journal of Sex Research*, 34 (2), 1 January 1997: 167–74.

16. Mark, K. P., Janssen, E., Milhausen, R. R. Infidelity in heterosexual couples: Demographic, interpersonal, and personality-related predictors of extradyadic sex. *Archives of Sexual Behaviour*, 40 (5), 1 October 2011: 971–82.

17. Higgins, R. and Meredith, P. 'Ngā tamariki – Māori childhoods – Māori children's upbringing. *Te Ara: The Encyclopedia of New Zealand*, updated 23-May-13 Available from: http://www.TeAra.govt.nz/en/nga-tamariki-maori-childhoods/page-2.

18. Bellis, M. A., Hughes, K., Hughes, S., Ashton, J. R. Measuring Paternal Discrepancy and its public health consequences. *Journal of Epidemiology and Community Health*, 59 (9), 1 September 2005: 749–54.

19. Bellis et al. Measuring Paternal Discrepancy.

20. Wilson, B., Smallwood, S. The proportion of marriages ending in divorce. *Population Trends*, 13, 2008: 28.

21. Office for National Statistics. What Percentage of Marriages End in Divorce?

22. Office for National Statistics. What Percentage of Marriages End in Divorce? Available from: http://www.ons.gov.uk/ons/rel/vsob1/divorces-in-england-and-wales/2011/sty-what-percentage-of-marriages-end-in-divorce.html [accessed 5 September 2014].

23. Nemesis, S. Female Divorce Risk Calculator. Just Four Guys. Available from: http://www.justfourguys.com/female-divorce-risk-calculator/ [accessed 21 October 2014].

24. Wilson, B., Stuchbury, R. Do partnerships last? Comparing marriage and cohabitation using longitudinal census data. *Population Trends*, 139, Spring 2010. Available from: http://www.ons.gov.uk/ons/rel/population-trends-rd/population-trends/no--139--spring-2010/index.html [accessed 5 September 2014].

25. Office for National Statistics. Divorces in England and Wales, 2012. Available from: http://www.ons.gov.uk/ons/rel/vsob1/divorces-in-england-and-wales/2012/stb-divorces-2012.html [accessed 5 September 2014].

26. Owen, J., Fincham, F. D. Effects of gender and psychosocial factors on 'friends with benefits' relationships among young adults. *Archives of Sexual Behavior*, 40 (2), 1 April 2011: 311–20.

27. Bisson, M. A., Levine, T. R. Negotiating a friends with benefits relationship. *Archives of Sexual Behavior*, 38 (1), 1 February 2009: 66–73.

4: Activities with the Opposite Sex

1. Brillinger, D. R. John W. Tukey: His life and professional contributions. *Annals of Statistics*, 30 (6), December 2002: 1535–75.

2. Brillinger, John W. Tukey.

3. Kinsey, A. C., Pomeroy, W. B., Martin, C. E. *Sexual Behavior in the Human Male*. Indiana University Press, 1948.

4. Gebhard, P. H., Johnson, A. B., Kinsey, A. C. *The Kinsey Data: Marginal Tabulations of the 1938–1963 Interviews Conducted by the Institute for Sex Research*. Indiana University Press, 1979.

5. Kinsey Institute. Data & Codebooks [Research Program]. Available from: http://www.kinseyinstitute.org/research/kidata.html [accessed 4 September 2014].

6. Kinsey Institute. Data & Codebooks.

7. Kinsey Institute. Data & Codebooks.

8. Kinsey et al. *Sexual Behavior in the Human Male*.

9. Kinsey, A. C. *Sexual Behavior in the Human Female*. Indiana University Press, 1953.

10. Mercer, C. H., Tanton, C., Peah, P., Erens, B., Sonnenberg, P., Clifton, S., et al. Changes in sexual attitudes and lifestyles in Britain through the life course and over time: Findings from the National Surveys of Sexual Attitudes and Lifestyles (Natsal). *Lancet*, 382 (9907), 30 November 2013: 1781–94.

11. Linda, S. T., Schumm, L. P., Laumann, E. O., Levinson, W., O'Muircheartaigh, C. A., Waite, L. J. A study of sexuality and health among older adults in the United States. *New England Journal of Medicine*, 357 (8), 23 August 2007: 762–74.

12. Kinsey et al. *Sexual Behavior in the Human Male*.

13. Kinsey et al. *Sexual Behavior in the Human Male*.

14. Elliot L. *Sex on Campus: The Details Guide to the Real Sex Lives of College Students*. Random House, 1997.

15. Kinsey et al. *Sexual Behavior in the Human Male*.

16. Simon, W., Kraft, D., Kaplan, H. Oral sex: A critical overview. *AIDS and Sex: An Integrated Biomedical and Behavioural Approach*. Oxford University Press, 1990.

17. Neret, G. *Erotica Universalis*. Taschen Books, 1994.

18. Van der Velde, T. H. *Ideal Marriage, Its Physiology and Technique*. 1928. Available from: http://www.goodreads.com/book/show/1200322. Ideal_Marriage_Its_Physiology_and_Technique [accessed 4 September 2014].
19. Szreter, S., Fisher, K. *Sex before the Sexual Revolution: Intimate Life in England, 1918–1963*. Cambridge University Press, 2010.
20. Brooks, X. Michael Douglas on Liberace, Cannes, cancer and cunnilingus. *The Guardian Online*. Available from: http://www.theguardian.com/film/2013/jun/02/michael-douglas-liberace-cancer-cunnilingus?guni=Article:in%20body%20link [accessed 14 October 2014].
21. Owen, D. H., Katz, D. F. A review of the physical and chemical properties of human semen and the formulation of a semen simulant. *Journal of Andrology*, 26 (4), 8 July 2005: 459–69.
22. Morrissey, T. E. A complete breakdown of the nutritional content of semen. *Jezebel*. Available from: http://jezebel.com/a-complete-breakdown-of-the-nutritional-content-of-semen-953356816 [accessed 11 October 2014].
23. McBride, K. R., Fortenberry, J. D. Heterosexual anal sexuality and anal sex behaviors: A review. *Journal of Sex Research*, 47 (2–3), 24 March 2010: 123–36.
24. Gebhard et al. *The Kinsey Data*.
25. Hite, S. *The Hite Report*.
26. Wolfe, L. *The Cosmo Report*. Arbor House, 1981.
27. Hunt, M. M. *Sexual Behavior in the 70s*. Playboy Press, 1974.
28. Michael, R. T., Gagnon, J. H., Laumann, E. O., Kolata, G. *Sex in America: A Definitive Survey*. Little, Brown & Co., 1994.
29. Mercer et al. Changes in sexual attitudes and lifestyles in Britain.
30. Chandra, A., Copen, C. E., Mosher, W. D. Sexual behavior, sexual attraction, and sexual identity in the United States: Data from the 2006–2010 National Survey of Family Growth. *International Handbook on the Demography of Sexuality*. Springer, 2013, pp. 45–66.
31. McBride and Fortenberry. Heterosexual anal sexuality.
32. Sandnabba, N. K., Santtila, P., Alison, L., Nordling, N. Demographics, sexual behaviour, family background and abuse experiences of practitioners of sadomasochistic sex: A review of recent research. *Sex and Relationship Therapy*, 17 (1), 1 February 2002: 39–55.
33. Smith, A. M. A., Rissel, C. E., Richters, J., Grulich, A. E., de Visser, R. O. Sex in Australia: The rationale and methods of the Australian Study of Health and Relationships. *Australia and New Zealand Journal of Public Health*, 27 (2), 2003: 106–17.
34. Richters, J., de Visser, R. O., Rissel, C. E., Grulich, A. E., Smith, A. M. A. Demographic and psychosocial features of participants

in bondage and discipline, 'sadomasochism' or dominance and submission (BDSM): Data from a national survey. *Journal of Sexual Medicine*, 5 (7), 1 July 2008: 1660–68.

35. Wismeijer, A. A. J., van Assen, M. A. L. M. Psychological characteristics of BDSM practitioners. *Journal of Sexual Medicine*, 10 (8), 1 August 2013: 1943–52.

36. Hoxton Dungeon Suite. Packages. 2014. Available from: http://www.hoxtondungeonsuite.co.uk/packages/ [accessed 11 October 2014].

37. Millward, J. Down the Rabbit Hole: An Analysis of One Million Sex Toy Sales. Available from: http://jonmillward.com/blog/studies/down-the-rabbit-hole-analysis-1-million-sex-toy-sales/ [accessed 21 October 2014].

38. Huntingdon Health Promotion and HIV Support Team. *69 Ideas to Help You Enjoy Safer Sex*. Huntingdonshire Community Health Services, 1989.

39. *Glamour* magazine. 50 Amazing Sex Facts You Never Knew (Promise). Available from: http://www.glamourmagazine.co.uk/features/relationships/2009/10/02/50-sex-facts-you-never-knew [accessed 4 September 2014].

40. Heiman, J. R., Long, J. S., Smith, S. N., Fisher, W. A., Sand, M. S., Rosen, R. C. Sexual satisfaction and relationship happiness in midlife and older couples in five countries. *Archives of Sexual Behavior*, 40 (4), 1 August 2011: 741–53.

41. Hunt. *Sexual Behavior in the 70s*.

42. Cochran, W. G., Mosteller, F., Tukey, J. W. Statistical problems of the Kinsey Report. *Journal of the American Statistical Association*, 48 (264), December 1953: 673.

43. Kinsey et al. *Sexual Behavior in the Human Male*.

44. Ericksen, J. A., Steffen, S. A. *Kiss and Tell: Surveying Sex in the Twentieth Century*. Harvard University Press, 2009.

45. Jones, J. H. *Alfred C. Kinsey: A Life*. W. W. Norton & Co., 2004.

46. Ericksen, J. A. With enough cases, why do you need statistics? Revisiting Kinsey's methodology. *Journal of Sex Research*, 35 (2), 1 May 1998: 132–40.

47. Gebhard et al. *The Kinsey Data*.

5: Activities between People of the Same Sex

1. Wolff, C. *Magnus Hirschfeld: A Portrait of a Pioneer in Sexology*. Quartet Books, 1986.

2. Bullough, V. L. *Science in the Bedroom: A History of Sex Research*. Basic Books, 1994.

3. Isherwood, C. *Christopher and His Kind*. Random House, 2012.
4. Ford, C. S., Beach, F. A. *Patterns of Sexual Behavior*. Harper & Row, 1972.
5. Foucault, M. *The History of Sexuality: An Introduction*. Knopf Doubleday, 2012.
6. Kinsey, A. C., Pomeroy, W. B., Martin, C. E. *Sexual Behavior in the Human Male*. Indiana University Press, 1948.
7. Kinsey, A. C. *Sexual Behavior in the Human Female*. Indiana University Press, 1953.
8. Voeller, B. Some uses and abuses of the Kinsey scale. *Homosexuality, Heterosexuality: Concepts of Sexual Orientation*, 1990: 35.
9. Voeller, B. Some uses and abuses of the Kinsey scale.
10. Voeller, B. Some uses and abuses of the Kinsey scale.
11. Billy, J. O., Tanfer, K., Grady, W. R., Klepinger, D. H. The sexual behavior of men in the United States. *Family Planning Perspectives*, 25 (2), April 1993: 52–60.
12. Michael, R. T., Gagnon, J. H., Laumann, E. O., Kolata. G. *Sex in America: A Definitive Survey*. Little, Brown & Co. 1994.
13. Mercer, C. H., Tanton, C., Prah, P., Erens, B., Sonnenberg, P., Clifton, S., et al. Changes in sexual attitudes and lifestyles in Britain through the life course and over time: Findings from the National Surveys of Sexual Attitudes and Lifestyles (Natsal). *Lancet*, 382 (9907), 30 November 2013: 1781–94.
14. Office for National Statistics. Sexual Identity in the UK. Available from: http://www.ons.gov.uk/ons/rel/integrated-household-survey/integrated-household-survey/january-to-december-2012/info-sexual-identity.html [accessed 4 September 2014].
15. Gates, G. J. How Many People are Lesbian, Gay, Bisexual and Transgender? Williams Institute. Available from: http://williamsinstitute.law.ucla.edu/research/census-lgbt-demographics-studies/how-many-people-are-lesbian-gay-bisexual-and-transgender/ [accessed 4 September 2014].
16. Ward, B. W., Dahlhamer, J. M., Galinsky, A. M., Joestl, S. S. Sexual orientation and health among US adults: National Health Interview Survey, 2013. *National Health Statistical Reports*, (77), 2014: 1–12.
17. *New York Times*. How Many Americans Are Lesbian, Gay or Bisexual? Available from: http://well.blogs.nytimes.com/2014/07/21/how-many-americans-are-lesbian-gay-or-bisexual/ [accessed 5 October 2014].
18. Copas, A. J., Wellings, K., Erens, B., Mercer, C. H., McManus, S., Fenton, K. A., et al. The accuracy of reported sensitive sexual behaviour in Britain: Exploring the extent of change, 1990–2000. *Sexually Transmitted Infections*, 78 (1), 1 February 2002: 26–30.

19. Prah, P., Copas, A. J., Mercer, C. H., Clifton, S., Erens, B., Phelps, A., et al. Consistency in reporting sensitive sexual behaviours in Britain: Change in reporting bias in the second and third National Surveys of Sexual Attitudes and Lifestyles (Natsal-2 and Natsal-3). *Sexually Transmitted Infections*, 90 (2), March 2014: 90–93.
20. Wellings, K., Field, J., Johnson, A. M., Wadsworth, J. *Sexual Behaviour in Britain: The National Survey of Sexual Attitudes and Lifestyles*. Penguin Books, 1994.
21. Mercer et al. Changes in sexual attitudes and lifestyles in Britain through the life course and over time: Findings from the National Surveys of Sexual Attitudes and Lifestyles (Natsal).
22. Chandra, A., Copen, C. E., Mosher, W. D. Sexual behavior, sexual attraction, and sexual identity in the United States: Data from the 2006–2010 National Survey of Family Growth. *International Handbook on the Demography of Sexuality*. Springer, 2013: 45–66.
23. Blair, J. A. Probability sample of gay urban males: The use of two-phase adaptive sampling. *Journal of Sex Research*, 36 (1), 1 February 1999: 39–44.
24. Rosenberger, J. G., Reece, M., Schick, V., Herbenick, D., Novak, D. S, Van der Pol, B., et al. Sexual behaviors and situational characteristics of most recent male-partnered sexual event among gay and bisexually identified men in the United States. *Journal of Sexual Medicine*, 8 (11), 1 November 2011: 3040–50.
25. Bailey, J. V., Farquhar, C., Owen, C., Whittaker, D. Sexual behaviour of lesbians and bisexual women. *Sexually Transmitted Infections*, 79 (2), 1 April 2003: 147–50.
26. Mercer, C. H., Bailey, J. V., Johnson, A. M., Erens, B., Wellings, K., Fenton, K. A., et al. Women who report having sex with women: British national probability data on prevalence, sexual behaviors, and health outcomes. *American Journal of Public Health*, 97 (6), 1 June 2007: 1126–33.
27. Maier, T. Can Psychiatrists Really 'Cure' Homosexuality? Available from: http://www.scientificamerican.com/article/homosexuality-cure-masters-johnson/ [accessed 18 October 2014].
28. Kallmann, F. J. Twin and sibship study of overt male homosexuality. *American Journal of Human Genetics*, 4 (2), June 1952: 136–46.
29. Whitam, F. L., Diamond, M., Martin, J. Homosexual orientation in twins: A report on 61 pairs and three triplet sets. *Archives of Sexual Behavior*, 22 (3), 1 June 1993: 187–206.
30. Wiederman, M. W. *Understanding Sexuality Research*. Wadsworth, 2001.
31. Bailey, J. M., Dunne, M. P., Martin, N. G. Genetic and environmental influences on sexual orientation and its correlates in an Australian

twin sample. *Journal of Personal and Social Psychology*, 78 (3), 2000: 524–36.

32. Whitam et al. Homosexual orientation in twins. Eckert, E. D., Bouchard, T. J., Bohlen, J., Heston, L. L. Homosexuality in monozygotic twins reared apart. *British Journal of Psychiatry*, 148 (4), 1 April 1986: 421–5.

33. Hines, M. Gendered development. In: Lerner R. M., Lamb, M. E. (eds). *Handbook of Child Development and Developmental Science*, 7th edn. John Wiley & Sons, 2015.

34. Reimers, S. The BBC internet study: General methodology. *Archives of Sexual Behavior*, 36 (2), 1 April 2007: 147–61.

35. Manning, J. T., Fink, B. Digit ratio (2D:4D), dominance, reproductive success, asymmetry, and sociosexuality in the BBC internet study. *American Journal of Human Biology*, 20 (4), 1 July 2008: 451–61.

36. Grimbos, T., Dawood, K., Burriss, R. P., Zucker, K. J, Puts, D. A. Sexual orientation and the second to fourth finger length ratio: A meta-analysis in men and women. *Behavioural Neurosciences*, 124 (2), April 2010: 278–87.

37. Coates, J. M., Gurnell, M., Rustichini, A. Second-to-fourth digit ratio predicts success among high-frequency financial traders. *Proceedings of the National Academy of Sciences*, 106 (2), 13 January 2009: 623–8.

38. Schwerdtfeger, A., Heims, R., Heer, J. Digit ratio (2D:4D) is associated with traffic violations for male frequent car drivers. *Accident Analysis and Prevention*, 42 (1), January 2010: 269–74.

39. Collaer, M. L., Reimers, S., Manning, J. T. Visuospatial performance on an internet line judgment task and potential hormonal markers: Sex, sexual orientation, and 2D:4D. *Archives of Sexual Behavior*, 36 (2), 1 April 2007: 177–92.

40. Blanchard, R., Bogaert, A. F. Homosexuality in men and number of older brothers. *American Journal of Psychiatry*, 153 (1), January 1996: 27–31.

41. Blanchard, R. Fraternal birth order and the maternal immune hypothesis of male homosexuality. *Hormones and Behaviour*, 40 (2), September 2001: 105–14.

42. Lalumière, M. L., Blanchard, R., Zucker, K. J. Sexual orientation and handedness in men and women: A meta-analysis. *Psychological Bulletin*, 126 (4), July 2000: 575–92.

43. Blanchard, R., Lippa, R. A. Birth order, sibling sex ratio, handedness, and sexual orientation of male and female participants in a BBC internet research project. *Archives of Sexual Behavior*, 36 (2), 1 April 2007: 163–76.

44. Gathorne-Hardy, J. *Alfred C. Kinsey: Sex the Measure of All Things*. Pimlico, 1999.

6: By Your Own Hand

1.   Laqueur, T. W. *Solitary Sex: A Cultural History of Masturbation*. Zone Books, 2003.

2.   Pepys, S. *Diary*. Available from: http://www.pepys.info/ [accessed 4 September 2014].

3.   Marten, J. *Onania, or, The Heinous Sin of Self-Pollution, and All Its Frightfull Consequences in Both Sexes*. 1723.

4.   Taylor, B. Too much. *London Review of Books*, 6 May 2004: 22–4.

5.   Tissot, S. A. D. *Onanism: or, A Treatise upon the Disorders Produced by Masturbation: or, The Dangerous Effects of Secret and Excessive Venery*, trans. A. Hume. Ecco Print Editions, 1766.

6.   Acton, W. *The Functions and Disorders of the Reproductive Organs*. Lindsay and Blakiston, 1857.

7.   Bullough, V. L. *Science in the Bedroom: A History of Sex Research*. Basic Books, 1994.

8.   Ellis, H. *Studies in the Psychology of Sex*, vol. 1, *The Evolution of Modesty; The Phenomena of Sexual Periodicity; Auto-Erotism*. Available from: http://www.gutenberg.org/ebooks/13610 [accessed 5 October 2014].

9.   Hunt, A. The great masturbation panic and the discourses of moral regulation in nineteenth- and early twentieth-century Britain. *Journal of the History of Sexuality*, 1998: 575–615.

10. Brockman, F. S. A study of the moral and religious life of 251 preparatory school students in the United States. *The Pedagogical Seminary*, 9 (3), 1 September 1902: 255–73. Ericksen, J. A., Steffen, S. A. *Kiss and Tell: Surveying Sex in the Twentieth Century*. Harvard University Press, 2009.

11. Laqueur, *Solitary Sex*.

12. The Quack Doctor. La Vida Vibrator. Available from: http://thequackdoctor.com/index.php/la-vida-vibrator/ [accessed 30 October 2014].

13. Gebhard, P. H., Johnson, A. B., Kinsey, A. C. *The Kinsey Data: Marginal Tabulations of the 1938–1963 Interviews Conducted by the Institute for Sex Research*. Indiana University Press, 1979.

14. Ellis, *Studies in the Psychology of Sex*, vol. 1, *The Evolution of Modesty*.

15. Bullough, *Science in the Bedroom*.

16. Szreter, S., Fisher, K. *Sex before the Sexual Revolution: Intimate Life in England 1918–1963*. Cambridge University Press, 2010.

17. Davis, K. *Factors in the Sex Life of Twenty-Two Hundred Women*. Harper and Brothers, 1929.

18. Jones, J. H. *Alfred C. Kinsey: A Life*. W. W. Norton & Co., 2004.

19. Kinsey, A. C., Pomeroy, W. B., Martin, C. E. *Sexual Behavior in the Human Male*. Indiana University Press, 1948.

20. Gebhard, *The Kinsey Data*.
21. Jones, *Alfred C. Kinsey: A Life*.
22. Kinsey, A. C. *Sexual Behavior in the Human Female*. Indiana University Press, 1953.
23. Masters, W. H., Johnson, V. E. *Human Sexual Response*. Ishi Press International, 2010.
24. Hite, S. *The Hite Report: On Female Sexuality*. Pandora, 1976.
25. Wolfe, L. *The Cosmo Report*. Arbor House, 1981.
26. Michael, R. T., Gagnon, J. H., Laumann, E. O., Kolata. G. *Sex in America: A Definitive Survey*. Little, Brown & Co., 1994.
27. Wellings, K., Field, J., Johnson, A. M., Wadsworth, J. *Sexual Behaviour in Britain: The National Survey of Sexual Attitudes and Lifestyles*. Penguin Books, 1994.
28. Gerressu, M., Mercer, C. H., Graham, C. A., Wellings, K., Johnson. A. M. Prevalence of masturbation and associated factors in a British national probability survey. *Archives of Sexual Behavior*, 37 (2), 1 April 2008: 266–78.
29. Mercer, C. H., Tanton, C., Prah, P., Erens, B., Sonnenberg, P., Clifton, S., et al. Changes in sexual attitudes and lifestyles in Britain through the life course and over time: findings from the National Surveys of Sexual Attitudes and Lifestyles (Natsal). *Lancet*, 382 (9907), 30 November 2013: 1781–94.
30. Gerressu et al. Prevalence of masturbation and associated factors in a British national probability survey.
31. Herbenick, D., Reece, M., Sanders, S., Dodge, B., Ghassemi, A., Fortenberry, J. D. Prevalence and characteristics of vibrator use by women in the United States: Results from a nationally representative study. *Journal of Sexual Medicine*, 6 (7), 1 July 2009: 1857–66.
32. Millward, J. Down the Rabbit Hole: An Analysis of One Million Sex Toy Sales. Available from: http://jonmillward.com/blog/studies/down-the-rabbit-hole-analysis-1-million-sex-toy-sales/ [accessed 21 October 2014].
33. *Mail Online*. NHS recommends pupils have an 'orgasm a day' to reduce risk of heart attack and stroke. Available from: http://www.dailymail.co.uk/news/article-1199132/NHS-recommends-pupils-orgasm-day-reduce-risk-heart-attack-stroke.html [accessed 4 September 2014].
34. NoFap, Partake in the Ultimate Challenge. NoFap. Available from: http://www.nofap.org/ [accessed 4 September 2014].
35. Voon, V., Mole, T. B., Banca, P., Porter, L., Morris, L., Mitchell, S., et al. Neural correlates of sexual cue reactivity in individuals with and without compulsive sexual behaviours. *PLoS ONE*, 9 (7), 11 June 2014: e102419.

36. *Mail Online*. Regular porn users show same brain activity as drug addicts. Available from: http://www.dailymail.co.uk/news/article-2428861/Porn-addicts-brain-activity-alcoholics-drug-addicts.html [accessed 21 October 2014].

7: How It All Starts

1. Copas, A. J., Wellings, K., Erens, B., Mercer, C. H., McManus, S., Fenton, K. A., et al. The accuracy of reported sensitive sexual behaviour in Britain: Exploring the extent of change 1990–2000. *Sexually Transmitted Infections*, 78 (1), 1 February 2002: 26–30. Prah, P., Copas, A. J., Mercer, C. H., Clifton, S., Erens, B., Phelps, A., et al. Consistency in reporting sensitive sexual behaviours in Britain: change in reporting bias in the second and third National Surveys of Sexual Attitudes and Lifestyles (Natsal-2 and Natsal-3). *Sexually Transmitted Infections*, 90 (2), March 2014: 90–93.
2. Hurwitt, M. Michael Schofield obituary. *The Guardian*, 27 April 2014. Available from: http://www.theguardian.com/law/2014/apr/27/michael-schofield [accessed 5 September 2014].
3. Schofield, M., Bynner, J., Lewis, P., Massie, P. *The Sexual Behaviour of Young People*. Penguin Books, 1976.
4. Channel 4. Highlights from YouGov's 'Sex Education' Survey: Sexperience. Available from: http://sexperienceuk.channel4.com/teen-sex-survey [accessed 5 September 2014].
5. YouGov. Survey Report: Channel 4 Sex Survey Results. Available from: http://d25d2506sfb94s.cloudfront.net/today_uk_import/YG-Archives-lif-ch4-sexed-090910.pdf [accessed 5 September 2014].
6. Upchurch, D. M., Lillard, L. A., Aneshensel, C. S., Li, N. F. Inconsistencies in reporting the occurrence and timing of first intercourse among adolescents. *Journal of Sex Research*, 239 (3), 1 August 2002: 197–206.
7. Centers for Disease Control and Prevention. YRBSS: Youth Risk Behavior Surveillance System – Adolescent and School Health. Available from: http://www.cdc.gov/HealthyYouth/yrbs/index.htm [accessed 5 September 2014].
8. Centers for Disease Control and Prevention. Youth Risk Behavior Surveillance: United States, 2013. Morbidity and Mortality Weekly Report, vol. 63, no. 4. Available from: http://www.cdc.gov/mmwr/pdf/ss/ss6304.pdf [accessed 5 September 2014].
9. Copen, C. E., Chandra, A., Martinez, G. Prevalence and Timing of Oral Sex with Opposite-Sex Partners among Females and Males Aged 15–24 Years: United States, 2007–2010. National Health

Statistics Reports, no. 56 (8/2012). Available from: http://www.cdc.gov/nchs/data/nhsr/nhsr056.pdf [accessed 5 September 2014].

10. Halpern-Felsher, B. L., Cornell, J. L., Kropp, R. Y., Tschann, J. M. Oral versus vaginal sex among adolescents: Perceptions, attitudes, and behavior. *Pediatrics*, 115 (4), 1 April 2005: 845–51.

11. Lifeway. True Love Waits. Available from: http://www.lifeway.com/n/Product-Family/True-Love-Waits [accessed 5 September 2014].

12. Brückner, H., Bearman, P. After the promise: The STD consequences of adolescent virginity pledges. *Journal of Adolescent Health*, 36 (4), April 2005: 271–8.

13. Rosenbaum, J. E. Patient Teenagers? A comparison of the sexual behavior of virginity pledgers and matched nonpledgers. *Pediatrics*, 123 (1), 1 January 2009: e110–20.

14. Culp-Ressler, T. Federal Funds Awarded to Abstinence-Only Education Programs. Available from: http://thinkprogress.org/health/2012/10/10/987411/federal-funds-abstinence-only-programs/ [accessed 5 September 2014].

15. HBSC. Health Behaviour in School-Aged Children. Available from: http://www.hbsc.org/ [accessed 5 September 2014].

16. Wellings, K., Nanchahal, K., Macdowall, W., McManus, S., Erens, B., Mercer, C. H., et al. Sexual behaviour in Britain: Early heterosexual experience. *Lancet*, 358 (9296), 1 December 2001: 1843–50.

17. Palmer, M. J., Clarke, L., Wellings, K. Is 'sexual competence' at first heterosexual intercourse associated with subsequent sexual health? Abstract. Available from: http://epc2014.princeton.edu/abstracts/140569 [accessed 5 September 2014].

18. Abma, J. C., Martinez, G., Copen, C. E. Teenagers in the United States: Sexual activity, contraceptive use, and childbearing, National Survey of Family Growth 2006–2008. *Vital Health Statistics*, 23 (30), June 2010: 1–47. Guttmacher Institute. American Teens' Sexual and Reproductive Health. Available from: http://www.guttmacher.org/pubs/FB-ATSRH.html [accessed 5 September 2014].

19. Sexual behaviour of young people. Editorial. *British Medical Journal*, 2 (5456), 31 July 1965: 247–9.

20. Martinez, G., Copen, C. E., Abma, J. C. Teenagers in the United States: Sexual activity, contraceptive use, and childbearing, 2006–2010: National Survey of Family Growth. *Vital Health Statistics*, 23 (31), October 2011: 1–35.

21. HBSC. Health Behaviour in School-Aged Children.

22. Office for National Statistics. Conceptions in England and Wales, 2012. Available from: http://www.ons.gov.uk/ons/rel/vsob1/

conception-statistics--england-and-wales/2012/2012-conceptions-statistical-bulletin.html [accessed 5 September 2014].

23. Office for National Statistics. Conceptions in England and Wales, 2012.

24. Bell, J., Clisby, S., Craig, G., Measor, L., Petrie, S., Stanley, N. Living on the Edge: Sexual Behaviour and Young Parenthood in Rural and Seaside Areas. University of Hull. Available from: https://www.education.gov.uk/publications/eOrderingDownload/RW8.pdf [accessed 2 October 2014].

25. Office for National Statistics. Conceptions in England and Wales, 2012.

26. United Nations Statistics Division. Millennium Indicators: 5.4. Adolescent Birth Rate per 1,000 Women. Available from: http://mdgs.un.org/unsd/mdg/Metadata.aspx?IndicatorId=0&SeriesId=761 [accessed 17 October 2014].

27. Wellings, K., Collumbien, M., Slaymaker, E., Singh, S., Hodges, Z., Patel, D., et al. Sexual behaviour in context: A global perspective. *Lancet*, 368 (9548), 17 November 2006: 1706–28.

8: Feelings about Sex

1. Degler, C. N. What ought to be and what was: Women's sexuality in the nineteenth century. *American Historical Review*, 79 (5), December 1974: 1467.

2. Platoni, K. The sex scholar. *Stanford Magazine*, 2010. Available from: http://alumni.stanford.edu/get/page/magazine/article/?article_id=29954 [accessed 5 September 2014].

3. Mosher C., Hygiene and Physiology of Women: Stanford Digital Repository Available from: http://purl.stanford.edu/sr010vc5273 [accessed 5 September 2014].

4. Mosher, C. D. *The Mosher Survey: Sexual Attitudes of 45 Victorian Women*. Arno Press, 1980.

5. Acton, W. *The Functions and Disorders of the Reproductive Organs*. Lindsay and Blakiston, 1857.

6. Cook, H. *The Long Sexual Revolution: English Women, Sex, and Contraception 1800–1975*. Oxford University Press, 2005.

7. McCance, R., Luff, M., Widdowson, E. Physical and emotional periodicity in women. *Journal of Hygiene*, 37 (04), 1937: 571–611.

8. Reproduced from McCance, Luff and Widdowson. Physical and emotional periodicity in women.

9. McCance, R. A., Widdowson, E. M. *McCance and Widdowson's The Composition of Foods*. Royal Society of Chemistry, 2002.

10. Szreter, S. *Fertility, Class and Gender in Britain, 1860–1940*. Cambridge University Press, 2002.

11. Caruso, S., Agnello, C., Malandrino, C., Lo Presti, L., Cicero, C., Cianci, S. Do hormones influence women's sex? Sexual activity over the menstrual cycle. *Journal of Sexual Medicine*, 11 (1), 1 January 2014: 211–21.

12. Regan, P. C. Rhythms of desire: The association between menstrual cycle phases and female sexual desire. *Canadian Journal of Human Sexuality*, 5, 1966: 145–56.

13. Michael, R. T., Gagnon, J. H., Laumann, E. O., Kolata, G. *Sex in America: A Definitive Survey*. Little, Brown & Co., 1994.

14. Fisher, T. D., Moore, Z. T., Pittenger, M.-J. Sex on the brain? An examination of frequency of sexual cognitions as a function of gender, erotophilia, and social desirability. *Journal of Sex Research*, 49 (1), 19 April 2011: 69–77.

15. Hofmann, W., Vohs, K. D., Baumeister, R. F. What people desire, feel conflicted about, and try to resist in everyday life. *Psychological Science*, 23 (6), 1 June 2012: 582–8.

16. Spector, I. P., Carey, M. P., Steinberg, L. The Sexual Desire Inventory: development, factor structure, and evidence of reliability. *Journal of Sexual and Marital Therapy*, 22 (3), 1996: 175–90.

17. Pittman, G.E. Who is Sir Francis Galton?. Available from: http:// www.galtoninstitute.org.uk/Newsletters/GINL0006/francis_galton. htm [accessed 5 September 2014].

18. Cooper-White, M. It's Better To Be Average – and 16 Other Surprising Laws of Human Sexual Attraction. Available from: http://www. huffingtonpost.com/2013/09/05/17-facts-about-human-sexual-attraction_n_3817941.html [accessed 5 September 2014].

19. Fletcher, G. J. O., Simpson, J. A., Thomas, G., Giles, L. Ideals in intimate relationships. *Journal of Personal and Social Psychology*, 76 (1), 1999: 72–89.

20. Singh, D. Adaptive significance of female physical attractiveness: Role of waist-to-hip ratio. *Journal of Personal and Social Psychology*, 65 (2), 1993: 293–307.

21. Streeter, S. A., McBurney, D. H. Waist–hip ratio and attractiveness: New evidence and a critique of 'a critical test'. *Evolution and Human Behaviour*, 24 (2), March 2003: 88–98.

22. Danamo's Marilyn Monroe Pages. Marilyn Monroe – Facts & Info. Available from: http://www.marilynmonroepages.com/facts/ [accessed 4 October 2014].

23. Healthy Celeb. Jessica Alba Height Weight Body Statistics. Available from: http://healthyceleb.com/jessica-alba-height-weight-body-statistics/3143 [accessed 4 October 2014].

24. Regan, Rhythms of desire.
25. Domingue, B. W., Fletcher, J., Conley, D., Boardman, J. D. Genetic and educational assortative mating among US adults. *Proceedings of the National Academy of Sciences*, 111 (22), 3 June 2014: 7996–8000.
26. Roach, M. *Bonk: The Curious Coupling of Sex and Science*. Canongate Books, 2008.
27. Plaud, J. J., Gaither, G. A., Hegstad, H. J., Rowan, L., Devitt, M. K. Volunteer bias in human psychophysiological sexual arousal research: To whom do our research results apply? *Journal of Sex Research*, 36 (2), 1 May 1999: 171–9.
28. Bloemers, J., Gerritsen, J., Bults, R., Koppeschaar, H., Everaerd, W., Olivier, B., et al. Induction of sexual arousal in women under conditions of institutional and ambulatory laboratory circumstances: a comparative study. *Journal of Sexual Medicine*, 7 (3), March 2010: 1160–76.
29. Woodard, T. L., Diamond, M. P. Physiologic measures of sexual function in women: A review. *Fertility and Sterility*, 92 (1), July 2009: 19–34.
30. Borg, C., De Jong, P. J. Feelings of disgust and disgust-induced avoidance weaken following induced sexual arousal in women. *PLoS ONE*, 7 (9), 12 September 2012: e44111.
31. England, L. R. Little Kinsey: An outline of sex attitudes in Britain. *Public Opinion Quarterly*, 13 (4), 21 December 1949: 587–600.
32. Stanley, L. *Sex Surveyed, 1949–1994: From Mass-Observation's 'Little Kinsey' to the National Survey and the Hite Reports*. Taylor & Francis, 1995.
33. Pew Research Center's Global Attitudes Project. Global Views on Morality. Available from: http://www.pewglobal.org/2014/04/15/global-morality/ [accessed 17 October 2014].
34. Platoni, K. The sex scholar.

9: Together at Last: Becoming a Couple

1. Szreter, S., Fisher, K. *Sex before the Sexual Revolution: Intimate Life in England 1918–1963*. Cambridge University Press, 2010.
2. Dunnell, K. *Family Formation 1976*. Her Majesty's Stationery Office, 1979. Available from: http://www.popline.org/node/448624 [accessed 16 October 2014].
3. Pew Research Center's Global Attitudes Project. Global Views on Morality. Available from: http://www.pewglobal.org/2014/04/15/global-morality/ [accessed 17 October 2014].
4. Wrigley, E. A., Schofield, R. S. *The Population History of England 1541–1871*. Cambridge University Press, 1989.

5. Wrigley, E. A. *English Population History from Family Reconstitution, 1580–1837.* Cambridge University Press, 1997.
6. Ruggles, S. The limitations of English family reconstitution: English population history from family reconstitution 1580–1837. *Continuity and Change,* 14 (1), May 1999: 105–30.
7. Wrigley, *English Population History from Family Reconstitution.*
8. Hair, P. E. Bridal pregnancy in earlier rural England further examined. *Population Studies,* 24 (1), March 1970: 59–70.
9. Hitchcock, T. Sex and gender: Redefining sex in eighteenth-century England. *History Workshop Journal,* 1996 (41), 1 January 1996: 72–90.
10. Registrar-General. *The Registrar-General's Statistical Review of England and Wales for the Years 1938 and 1939.* His Majesty's Stationery Office, 1947.
11. Szreter and Fisher, *Sex before the Sexual Revolution.*
12. Wrigley, *English Population History from Family Reconstitution.*
13. Office of Population Censuses and Surveys. *Birth Statistics: Historical Series 1837–1983. England and Wales.* Series FM1, no 13. Her Majesty's Stationery Office, 1983.
14. Wrigley, *English Population History from Family Reconstitution.*
15. Laslett, P., Oosterveen, K. Long-term trends in bastardy in England. *Population Studies,* 27 (2), 1 July 1973: 255–86.
16. Fletcher, J. Moral and educational statistics of England and Wales. *Journal of the Statistical Society of London,* 12 (3), 1 August 1849: 189–335.
17. Office for National Statistics. Births in England and Wales, 2012. Available from: http://www.ons.gov.uk/ons/rel/vsob1/birth-summary-tables--england-and-wales/2012/stb-births-in-england-and-wales-2012.html#tab-Live-births-within-marriage-civil-partnership [accessed 5 September 2014].
18. OECD Family Database. Births outside Marriage and Teenage Births Jan 2013. Available from: http://www.oecd.org/els/family/SF2_4_Births_outside_marriage_and_teenage_births_Jan2013.pdf [accessed 16 October 2014].
19. Office for National Statistics. Marriages in England and Wales (Provisional), 2012. Available from: http://www.ons.gov.uk/ons/rel/vsob1/marriages-in-england-and-wales--provisional-/2012/stb-marriages-in-england-and-wales--provisional---2011.html [accessed 5 September 2014].
20. *International Business Times UK.* Playboy model Cathy Schmitz, 24, marries Austrian billionaire, 81. Available from: http://www.ibtimes.co.uk/austrian-billionaire-marries-playboy-model-cathy-schmitz-celebrity-couples-huge-age-gaps-1465721 [accessed 17 September 2014].

21. Wilson, B., Smallwood, S. Age differences at marriage and divorce. *Population Trends*, 132, December 2007: 17–25.
22. Beaujouan, É., Ní Bhrolcháin, M. Cohabitation and marriage in Britain since the 1970s. *Population Trends*, 145 (1), September 2011: 35–59.
23. Wrigley, *English Population History from Family Reconstitution*.
24. Office for National Statistics. Marriages in England and Wales (Provisional), 2012
25. Office of Population Censuses and Surveys. *Birth Statistics: Historical Series 1837–1983. England and Wales*. Series FM1, no 13. Her Majesty's Stationery Office, 1983.
26. Office for National Statistics. Births in England and Wales, 2012.

10: Sex and Not Having Babies

1. Youssef, H. The history of the condom. *Journal of the Royal Society of Medicine*, 86 (4), April 1993: 226–8.
2. Szreter, S. *Fertility, Class and Gender in Britain, 1860–1940*. Cambridge University Press, 2002.
3. Alcott, W. A. *The Physiology of Marriage*. J. P. Jewett & Co., 1856.
4. Cook, H. *The Long Sexual Revolution: English Women, Sex, and Contraception 1800–1975*. Oxford University Press, 2005.
5. Brown, J. W., Greenwood, M., Wood, F. The fertility of the English middle classes: A statistical study. *Eugenics Review*, 12 (3), October 1920: 158–211.
6. Szreter, *Fertility, Class and Gender in Britain, 1860–1940*.
7. Lewis-Faning, E. *Report on an Enquiry into Family Limitation and Its Influence on Human Fertility during the Past Fifty Years*. His Majesty's Stationery Office, 1949.
8. Szreter, S., Fisher, K. *Sex before the Sexual Revolution: Intimate Life in England 1918–1963*. Cambridge University Press, 2010.
9. Szreter, *Fertility, Class and Gender in Britain, 1860–1940*.
10. Maisel, A. Q. *The Hormone Quest*. Random House, 1965.
11. Cook, *The Long Sexual Revolution*.
12. Furedi, A. Social consequences: The public health implications of the 1995 'pill scare'. *Human Reproduction Update*, 5 (6), 1 November 1999: 621–6.
13. BBC News. U-turn over Pill Scare. 7 April 1999. Available from: http://news.bbc.co.uk/1/hi/health/313848.stm [accessed 18 September 2014].
14. Robinson, C., Nardone, A., Mercer, C., Johnson, A. M. Health Survey for England 2010, Chapter 6, Sexual Health. Available from: http://www.hscic.gov.uk/catalogue/PUB03023/

heal-surv-eng-2010-resp-heal-ch6-sex.pdf [accessed 18 September 2014].

15. Eisenberg, D., McNicholas, C., Peipert, J. F. Cost as a barrier to long-acting reversible contraceptive (LARC) use in adolescents. *Journal of Adolescent Health*, 52 (4), supplement, April 2013: S59–63.

16. Trussell, J., Henry, N., Hassan, F., Prezioso, A., Law, A., Filonenko, A. Burden of unintended pregnancy in the United States: Potential savings with increased use of long-acting reversible contraception. *Contraception*, 87 (2), February 2013: 154–61.

17. NHS Choices. How Effective Is Contraception? Contraception Guide. Available from: http://www.nhs.uk/Conditions/contraception-guide/Pages/how-effective-contraception.aspx [accessed 5 September 2014].

18. Trussell, J. Contraceptive failure in the United States. *Contraception*, 83 (5), May 2011: 397–404.

19. Walsh, T. L., Frezieres, R. G., Peacock, K., Nelson, A. L., Clark, V. A., Bernstein, L., et al. Effectiveness of the male latex condom: Combined results for three popular condom brands used as controls in randomized clinical trials. *Contraception*, 70 (5), November 2004: 407–13.

20. Kost, K., Singh, S., Vaughan, B., Trussell, J., Bankole, A. Estimates of contraceptive failure from the 2002 National Survey of Family Growth. *Contraception*, 77 (1), January 2008: 10–21.

21. Killick, S. R., Leary, C., Trussell, J., Guthrie, K. A. Sperm content of pre-ejaculatory fluid. *Human Fertility*, 14 (1), March 2011: 48–52.

22. Whitby, A. Averages Deceive: Birth Control Is Better than the NYT Credits. Available from: http://andrewwhitby.com/2014/09/15/averages-deceive-birth-control-is-better-than-the-nyt-credits/ [accessed 20 October 2014].

23. Szreter, S. *Fertility, Class and Gender in Britain, 1860–1940*.

24. British Medical Association. *Report of Committee on Medical Aspects of Abortion*. British Medical Association, 1936.

25. Department of Health. Abortion Statistics, England and Wales. Available from: https://www.gov.uk/government/collections/abortion-statistics-for-england-and-wales [accessed 5 September 2014].

11: Sex and Having Babies

1. National Institute for Health and Care Excellence. Fertility, Introduction, Guidance and Guidelines. Available from: https://www.nice.org.uk/guidance/CG156/chapter/introduction [accessed 5 September 2014].

2. Twenge, J. M. *The Impatient Woman's Guide to Getting Pregnant*. Simon and Schuster, 2012. BBC. The 300-Year-Old Odds of Having a Baby. 18 September 2013. Available from: http://www.bbc.co.uk/news/magazine-24128176 [accessed 5 September 2014].

3. Heffner, L. J. Advanced maternal age: How old is too old? *New England Journal of Medicine*, 351 (19), 2004: 1927–9.

4. Heffner. Advanced maternal age. Menken, J., Trussell, J., Larsen, U. Age and infertility. *Science*, 233 (4771), 26 September 1986: 1389–94.

5. Twenge, J. How long can you wait to have a baby? *The Atlantic*, August 2013. Available from: http://www.theatlantic.com/magazine/archive/2013/07/how-long-can-you-wait-to-have-a-baby/309374/ [accessed 19 September 2014].

6. Wrigley, E. A. *English Population History from Family Reconstitution, 1580–1837*. Cambridge University Press, 1997.

7. Colombo, B., Masarotto, G. Daily fecundability. *Demographic Research*, 6 September 2000. Available from: http://www.demographic-research.org/Volumes/Vol3/5/default.htm [accessed 25 January 2012].

8. Dunson, D. B., Colombo, B., Baird, D. D. Changes with age in the level and duration of fertility in the menstrual cycle. *Human Reproduction*, 17 (5), 1 May 2002: 1399–403.

9. Dunson, D. B., Baird, D. D., Colombo, B. Increased infertility with age in men and women. *Obstetrics and Gynecology*, 103 (1), January 2004: 51–6.

10. National Institute for Health and Care Excellence. Fertility, Introduction, Guidance and Guidelines.

11. Oberzaucher, E., Grammer, K. The case of Moulay Ismael: Fact or fancy? *PLoS ONE*, 9 (2), 14 February 2014: e85292.

12. Mosher, W. D., Jones, J., Abma, J. C. *Intended and Unintended Births in the United States, 1982–2010*. US Department of Health and Human Services, Centers for Disease Control and Prevention, National Center for Health Statistics, 2012.

13. Wellings, K., Jones, K. G., Mercer, C. H., Tanton, C., Clifton, S., Datta, J., et al. The prevalence of unplanned pregnancy and associated factors in Britain: Findings from the third National Survey of Sexual Attitudes and Lifestyles (Natsal-3). *Lancet*, 382 (9907), 30 November 2013: 1807–16.

14. Wellings et al., The prevalence of unplanned pregnancy and associated factors in Britain.

15. Office for National Statistics. *Births in England and Wales by Characteristics of Birth 2, 2012*. Office for National Statistics, 2013. Available from: http://www.ons.gov.uk/ons/rel/

vsob1/characteristics-of-birth-2--england-and-wales/2012/
sb-characteristics-of-birth-2.html [accessed 5 September 2014].

16. *Mail Online*. Tis the season to conceive! December most likely time to
fall pregnant. Available from: http://www.dailymail.co.uk/femail/
article-2522048/Tis-season-conceive-December-common-month-
year-pregnant.html [accessed 5 September 2014].

17. Wrigley, E. A., Schofield, R. S. *The Population History of England
1541–1871*. Cambridge University Press, 1989.

18. Hair, P. E. Bridal pregnancy in earlier rural England further
examined. *Population Studies*, 24 (1), March 1970: 59–70.

19. Wellings, K., Macdowall, W., Catchpole, M., Goodrich, J. Seasonal
variations in sexual activity and their implications for sexual health
promotion. *Journal of the Royal Society of Medicine*, 92 (2), February
1999: 60–64.

20. *Mail Online*. Sandy baby boom begins! New Jersey hospitals report
that the number of deliveries are up by 30% nine months after the
hurricane. Available from: http://www.dailymail.co.uk/news/
article-2376861/Hurricane-Sandy-baby-boom-begins-New-Jersey-
hospitals-report-number-deliveries-30-months-later.html [accessed
21 September 2014].

21. CNN. Is the post-Sandy baby boom real? 2013. Available from:
http://www.cnn.com/2013/07/24/us/sandy-baby-boom/index.
html [accessed 21 September 2014]. *Science Daily*. Blackout baby
boom a myth, Duke professor says. Available from: http://www.
sciencedaily.com/releases/2004/05/040512044711.htm [accessed 21
September 2014].

22. BBC. Uganda Blackouts 'Fuel Baby Boom'. 12 March 2009. Available
from: http://news.bbc.co.uk/1/hi/world/africa/7939534.stm
[accessed 5 September 2014].

23. Burlando, A. Power outages, power externalities, and baby booms.
Demography, 51 (4), August 2014: 1477–500.

24. Fetzer, T., Pardo, O., Shanghavi, A. An Urban Legend?! Power
Rationing, Fertility and its Effects on Mothers. Centre for Economic
Performance, 2013, report no. dp1247. Available from: http://ideas.
repec.org/p/cep/cepdps/dp1247.html [21 September 2014].

25. Tachibana, C. Obama baby boom turns out to be a bust. Available
from: http://www.nbcnews.com/id/32286000/ns/health-
behavior/t/obama-baby-boom-turns-out-be-bust/ [accessed 5
September 2014].

26. Montesinos, J., Cortes, J., Arnau, A., Sanchez, J. A., Elmore, M.,
Macia, N., et al. Barcelona baby boom: Does sporting success affect
birth rate? *British Medical Journal*, 347, 17 December 2013: f7387–f7387.

27. Crawford, C., Dearden, L., Greaves, E. When You are Born Matters: Evidence for England. Institute of Fiscal Studies Report R80, 2013. Available from: http://www.ifs.org.uk/comms/r80.pdf [accessed 5 September 2014].

28. Rosenbaum, M. Birth Month Affects Oxbridge Chances. BBC. 2013. Available from: http://www.bbc.co.uk/news/uk-politics-21579484 [accessed 5 September 2014].

29. Crawford et al., When You are Born Matters.

30. Baker, J., Schorer, J., Cobley, S. Relative age effects. *Sportwissenschaft*, 40 (1), 1 March 2010: 26–30.

12: Pleasures and Problems

1. World Health Organization. Sexual Health. Available from: http://www.who.int/topics/sexual_health/en/ [accessed 5 November 2014].

2. Kinsey, A. C., Pomeroy, W. B., Martin, C. E. *Sexual Behavior in the Human Male*. Indiana University Press, 1948.

3. Wylie, K. R., Eardley, I. Penile size and the 'small penis syndrome'. *BJU International*, 99 (6), June 2007: 1449–55.

4. Lever, J., Frederick, D. A., Peplau, L. A. Does size matter? Men's and women's views on penis size across the lifespan. *Psychology of Men and Masculinity*, 7 (3), 2006: 129–43.

5. Barnhart, K. T., Izquierdo, A., Pretorius, E. S., Shera, D. M., Shabbout, M., Shaunik, A. Baseline dimensions of the human vagina. *Human Reproduction*, 21 (6), June 2006: 1618–22.

6. Giuliano, F., Patrick, D. L., Porst, H., La Pera, G., Kokoszka, A., Merchant, S., et al. Premature ejaculation: results from a five-country European observational study. *European Urology*, 53 (5), May 2008: 1048–57.

7. Smith, G. D., Frankel, S., Yarnell, J. Sex and death: are they related? Findings from the Caerphilly cohort study. *British Medical Journal*, 315 (7123), 20 December 1997: 1641–4.

8. Ebrahim, S., May, M., Shlomo, Y. B., McCarron, P., Frankel, S., Yarnell, J., et al. Sexual intercourse and risk of ischaemic stroke and coronary heart disease: the Caerphilly study. *Journal of Epidemiology and Community Health*, 56 (2), 1 February 2002: 99–102.

9. Casazza, K., Fontaine, K. R., Astrup, A., Birch, L. L., Brown, A. W., Bohan Brown, M. M., et al. Myths, presumptions, and facts about obesity. *New England Journal of Medicine*, 368 (5), 30 January 2013: 446–54.

10. Frappier, J., Toupin, I., Levy, J. J., Aubertin-Leheudre, M., Karelis, A. D. Energy expenditure during sexual activity in young healthy couples. *PLoS ONE*, 8 (10), 24 October 2013: e79342.

11. Dahabreh, I. J., Paulus, J. K. Association of episodic physical and sexual activity with triggering of acute cardiac events: Systematic review and meta-analysis. *Journal of the American Medical Association*, 305 (12), 23 March 2011: 1225–33.

12. Levine, G. N., Steinke, E. E., Bakaeen, F. G., Bozkurt, B., Cheitlin, M. D., Conti, J. B., et al. Sexual activity and cardiovascular disease: A scientific statement from the American Heart Association. *Circulation*, 125 (8), 28 February 2012: 1058–72.

13. Gallup, G. G. J., Burch, R. L., Platek, S. M. Does semen have antidepressant properties? *Archives of Sexual Behavior*, 31 (3), 1 June 2002: 289–93.

14. Peleg, R., Peleg, A. Case report: Sexual intercourse as potential treatment for intractable hiccups. *Canadian Family Physician*, 46, August 2000: 1631–2.

15. Herbenick, D., Schick, V., Reece, M., Sanders, S., Fortenberry, J. D. Pubic hair removal among women in the United States: Prevalence, methods, and characteristics. *Journal of Sexual Medicine*, 7 (10), 1 October 2010: 3322–30.

16. Millward, J. Down the Rabbit Hole: An Analysis of One Million Sex Toy Sales. Available from: http://jonmillward.com/blog/studies/down-the-rabbit-hole-analysis-1-million-sex-toy-sales/ [accessed 21 October 2014].

17. Michael, R. T., Gagnon, J. H., Laumann, E. O., Kolata, G. *Sex in America: A Definitive Survey*. Little, Brown & Co., 1994.

18. Layte, R., McGee, H., Quail, A., Rundle, K., Cousins, G., Donnelly, C., et al. The Irish Study of Sexual Health and Relationships Main Report. Royal College of Surgeons in Ireland Psychology Reports, 1 October 2006. Available from: http://epubs.rcsi.ie/psycholrep/35.

19. Ruiz-Muñoz, D., Wellings, K., Castellanos-Torres, E., Álvarez-Dardet, C., Casals-Cases, M., Pérez, G. Sexual health and socioeconomic-related factors in Spain. *Annals of Epidemiology*, 23 (10), October 2013: 620–28.

20. Laumann, E. O., Paik, A., Glasser, D. B., Kang, J.-H., Wang, T., Levinson, B., et al. A cross-national study of subjective sexual well-being among older women and men: Findings from the Global Study of Sexual Attitudes and Behaviors. *Archives of Sexual Behavior*, 35 (2), 1 April 2006: 143–59.

21. Mitchell, K. R., Mercer, C. H., Ploubidis, G. B., Jones, K. G., Datta, J., Field, N., et al. Sexual function in Britain: Findings from the third

National Survey of Sexual Attitudes and Lifestyles (Natsal-3). *Lancet*, 382 (9907), 30 November 2013: 1817–29.

22. Hite, S. *The Hite Report: On Female Sexuality*. Pandora, 1976.

23. Wallen, K., Lloyd, E. A. Female sexual arousal: Genital anatomy and orgasm in intercourse. *Hormones and Behaviour*, 59 (5), May 2011: 780–92.

24. Michael et al., *Sex in America: A Definitive Survey*.

25. Lindau, S. T., Schumm, L. P., Laumann, E. O., Levinson, W., O'Muircheartaigh, C. A., Waite, L. J. A study of sexuality and health among older adults in the United States. *New England Journal of Medicine*, 357 (8), 23 August 2007: 762–74.

26. Lindau et al., A study of sexuality and health among older adults in the United States.

27. Nicolosi, A., Laumann, E. O., Glasser, D. B., Moreira Jr, E. D., Paik, A., Gingell, C. Sexual behavior and sexual dysfunctions after age 40: The global study of sexual attitudes and behaviors. *Urology*, 64 (5), November 2004: 991–7.

28. Lewis, R. W., Fugl-Meyer, K. S., Corona, G., Hayes, R. D., Laumann, E. O., Moreira Jr, E. D., et al. ORIGINAL ARTICLES: Definitions/ Epidemiology/Risk Factors for Sexual Dysfunction. *Journal of Sexual Medicine*, 7 (4, pt 2), 1 April 2010: 1598–607.

29. IsHak, W., Tobia, G. DSM-5 changes in diagnostic criteria of sexual dysfunctions. *Reproductive System and Sexual Disorders*, 2, 2013: 122.

30. Anon. The hand of fate: Mr. A. Wilson's death on eve of thyroid lecture. *The Register* (Adelaide), 5 July 1921: 8.

31. Department of Health. NHS availability of erectile dysfunction drugs: proposed changes. Available from: https://www.gov.uk/ government/consultations/nhs-availability-of-erectile-dysfunction-drugs-proposed-changes [accessed 5 September 2014].

32. Venhuis, B. J., De Voogt, P., Emke, E., Causanilles, A., Keizers, P. H. J. Success of rogue online pharmacies: Sewage study of sildenafil in the Netherlands. *British Medical Journal*, 349, 2 July 2014: g4317–g4317.

33. Siegel-Itzkovich, J. Viagra makes flowers stand up straight. *British Medical Journal*, 319 (7205), 31 July 1999: 274.

13: Sex, Media and Technology

1. Reichert, T., Carpenter, C. An update on sex in magazine advertising: 1983 to 2003. *Journal of Mass Communication Quarterly*, 81 (4), 1 December 2004: 823–37.

2. Wyllie, J., Carlson, J., Rosenberger, P. J. Examining the influence of different levels of sexual-stimuli intensity by gender on advertising

effectiveness. *Journal of Marketing Management*, 30 (7–8), 8 January 2014: 697–718.

3. This is not advertising. Axe Detailer – Cleans Your Balls. 2011. Available from: http://thisisnotadvertising.wordpress. com/2011/07/11/axe-detailer-cleans-your-balls/ [accessed 6 September 2014].

4. Ofcom. UK Audience Attitudes to Broadcast Media. 2014. Available from: http://stakeholders.ofcom.org.uk/market-data-research/ other/tv-research/attitudes-to-broadcast-media/ [accessed 6 September 2014].

5. Brown, J. D., L'Engle, K. L., Pardun, C. J., Guo, G., Kenneavy, K., Jackson, C. Sexy media matter: Exposure to sexual content in music, movies, television, and magazines predicts black and white adolescents' sexual behavior. *Pediatrics*, 117 (4), 1 April 2006: 1018–27.

6. Steinberg, L., Monahan, K. C. Adolescents' exposure to sexy media does not hasten the initiation of sexual intercourse. *Developmental Psychology*, 47 (2), 2011: 562–76.

7. Magnanti, D. B. *The Sex Myth: Why Everything We're Told is Wrong.* Weidenfeld and Nicolson, 2012.

8. Covenant Eyes | The Leaders in Accountability Software. Porn Stats. 2014. Available from: http://www.covenanteyes.com/pornstats/ [accessed 6 September 2014].

9. Hald, G. M. Gender differences in pornography consumption among young heterosexual Danish adults. *Archives of Sexual Behavior*, 35 (5), 1 October 2006: 577–85.

10. Channel 4. Highlights from YouGov's 'Sex Education' Survey: Sexperience. Available from: http://sexperienceuk.channel4.com/ teen-sex-survey [accessed 5 September 2014].

11. Internet Adult Film Database. Available from: http://www.iafd.com [accessed 6 September 2014].

12. Millward, J. Deep Inside: A Study of 10,000 Porn Stars. 2014. Available from: http://jonmillward.com/blog/studies/deep-inside-a-study-of-10000-porn-stars/ [accessed 6 September 2014].

13. Peter, J., Valkenburg, P. M. The influence of sexually explicit internet material on sexual risk behavior: a comparison of adolescents and adults. *Journal of Health Communication*, 16 (7), August 2011: 750–65.

14. Zillmann, D., Bryant, J. Pornography's impact on sexual satisfaction. *Journal of Applied Social Psychology*, 18 (5), 1 April 1988: 438–53.

15. Morgan, E. M. Associations between young adults' use of sexually explicit materials and their sexual preferences, behaviors, and satisfaction. *Journal of Sex Research*, 48 (6), December 2011: 520–30.

16. Peter and Valkenburg. The influence of sexually explicit internet material on sexual risk behavior.

17. Zillmann, D. Influence of unrestrained access to erotica on adolescents' and young adults' dispositions toward sexuality. *Journal of Adolescent Health*, 27 (2), 1 August 2000: 41–4.

18. Lenhart, A. Teens and Sexting. Pew Research Center's Internet & American Life Project. Available from: http://www.pewinternet. org/2009/12/15/teens-and-sexting/ [accessed 6 September 2014].

19. Mitchell, K. J., Finkelhor, D., Jones, L. M., Wolak, J. Prevalence and characteristics of youth sexting: A national study. *Pediatrics*, 5 December 2011: peds. 2011–1730.

20. Strassberg, D. S., McKinnon, R. K., Sustaíta, M. A., Rullo, J. Sexting by high school students: An exploratory and descriptive study. *Archives of Sexual Behavior*, 42 (1), January 2013: 15–21.

21. National Campaign to Prevent Teen and Unplanned Pregnancy. Sex and Tech. 2008. Available from: http://thenationalcampaign.org/ resource/sex-and-tech [accessed 6 September 2014].

22. NSPCC. ChildLine Tackling Sexting with Internet Watch Foundation. Available from: http://www.nspcc.org.uk/news-and-views/media-centre/press-releases/2013/childline-internet-watch-foundation/ childline-tackling-sexting-internet-watch-foundation_wdn98995.html [accessed 6 September 2014].

23. BBC. Teenagers Face 'Sexting' Pressure. 16 October 2013. Available from: http://www.bbc.co.uk/news/uk-24539514 [accessed 6 September 2014].

24. NSPCC. A Qualitative Study of Children, Young People and 'Sexting'. 2012. Available from: http://www.nspcc.org.uk/Inform/ resourcesforprofessionals/sexualabuse/sexting-research_wda89260. html [accessed 6 September 2014].

25. ChildLine. Sexting: Online and Mobile Safety. 2014. Available from: http://www.childline.org.uk/explore/onlinesafety/pages/sexting. aspx [accessed 6 September 2014].

26. Papadopoulos, L. Sexualisation of Young People: A Review. 2010. Available from: http://webarchive.nationalarchives.gov.uk/+/http:/ www.homeoffice.gov.uk/documents/sexualisation-of-young-people. pdf [accessed 10 December 2014].

27. IPPR. Young People, Sex and Relationships: The New Norms. 2014. Available from: http://www.ippr.org/publications/young-people-sex-and-relationships-the-new-norms [accessed 6 September 2014].

28. YouGov. Survey Report: Channel 4 Sex Survey Results. Available from: http://d25d2506sfb94s.cloudfront.net/today_uk_import/ YG-Archives-lif-ch4-sexed-090910.pdf [accessed 5 September 2014].

29. Channel 4. Highlights from YouGov's 'Sex Education' Survey.

30. Hill, A. 'Streetwise' British teenagers are ignorant about sex, survey reveals. *The Guardian*, 7 September 2008. Available from: http://

www.theguardian.com/education/2008/sep/07/sexeducation.
youngpeople [accessed 25 September 2014].

31. Spiegelhalter, D. Numbers and the Common-Sense Bypass:
    Understanding Uncertainty. 2014. Available from: http://
    understanderuncertainty.org/numbers-and-common-sense-bypass
    [accessed 25 September 2014].

32. Jumio. Where Do You Take Your Phone? Available from: https://
    www.jumio.com/2013/07/where-do-you-take-your-phone/
    [accessed 10 November 2014].

33. Spreadsheets. Spreadsheets – #1 Sex App. Available from: http://
    spreadsheetsapp.com/ [accessed 10 November 2014].

14: The Dark Side: Prostitution, the Pox and Having Sex against Your Will

1. Acton, W. *Prostitution, Considered in Its Moral, Social & Sanitary Aspects, in London and Other Large Cities.* J. Churchill, 1857.

2. Colquhoun, P. A. *Treatise on the Police of the Metropolis: Explaining the Various Crimes and Misdemeanors Which at Present are Felt as a Pressure upon the Community; and Suggesting Remedies for Their Prevention.* H. Fry, 1796.

3. Mayhew, H., Tuckniss, W., Beeard, R. *London Labour and the London Poor: A Cyclopaedia of the Condition and Earnings of Those That Will Work, Those That Cannot Work, and Those That Will Not Work.* G. Woodfall and Son, 1862. Available from: http://archive.org/details/cu31924092592793 [accessed 7 September 2014].

4. Mason, M. *The Making of Victorian Sexuality.* Oxford University Press, 1994.

5. Acton, W. *The Functions and Disorders of the Reproductive Organs.* Lindsay and Blakiston, 1857.

6. Schei, B., Stigum, H. A study of men who pay for sex, based on the Norwegian National Sex Surveys. *Scandinavian Journal of Public Health*, 38 (2), 1 March 2010: 135–40.

7. Layte, R., McGee, H., Quail, A., Rundle, K., Cousins, G., Donnelly, C., et al. The Irish Study of Sexual Health and Relationships Main Report. Royal College of Surgeons in Ireland Psychology Reports, 1 October 2006. Available from: http://epubs.rcsi.ie/psycholrep/35 [10 December 2014].

8. Ward, H., Mercer, C. H., Wellings, K., Fenton, K., Erens, B., Copas, A., et al. Who pays for sex? An analysis of the increasing prevalence of female commercial sex contacts among men in Britain. *Sexually Transmitted Infections*, 81 (6), 1 December 2005: 467–71.

9. DHS Program. HIV/AIDS Survey Indicators Database: Indicators. 2014. Available from: http://hivdata.dhsprogram.com/ind_detl. cfm?ind_id=51&prog_area_id=8 [accessed 7 September 2014].

10. Kanouse, D. E., Berry, S. H., Duan, N., Lever, J., Carson, S., Perlman, J. F., et al. Drawing a probability sample of female street prostitutes in Los Angeles county. *Journal of Sex Research*, 36 (1), 1 February 1999: 45–51.

11. Potterat, J. J., Woodhouse, D. E., Muth, J. B., Muth, S. Q. Estimating the prevalence and career longevity of prostitute women. *Journal of Sex Research*, 27 (2), 1 May 1990: 233–43.

12. Office for National Statistics. Changes to National Accounts: Inclusion of Illegal Drugs and Prostitution in the UK National Accounts. Office for National Statistics. 2014. Available from: http:// www.ons.gov.uk/ons/rel/naa1-rd/national-accounts-articles/ inclusion-of-illegal-drugs-and-prostitution-in-the-uk-national-accounts/index.html [accessed 7 September 2014].

13. Magnanti, D. B. Prostitution 'adds £5bn a year to UK economy'. Are you having a laugh? *Daily Telegraph*, 30 May 2014. Available from: http://www.telegraph.co.uk/women/sex/10864898/Prostitution-adds-5bn-a-year-to-UK-economy.-Are-you-having-a-laugh.html [accessed 22 October 2014].

14. Jolyon, J. K. TaxRelief4escorts.. Available from: http://www. taxrelief4escorts.co.uk/ [accessed 22 October 2014].

15. Dickson, S. Mapping Commercial Sex across London. Poppy Project. 2008. Available from: http://i2.cmsfiles.com/eaves/2012/04/Sex-in-the-City-1751ff.pdf [accessed 7 September 2014].

16. Home Office. *Paying the Price: A Consultation Paper on Prostitution.* Home Office Communication Directorate. 2004. Available from: http://prostitution.procon.org/sourcefiles/paying_the_price.pdf. [accessed 10 December 2014].

17. Cusick, L., Kinnell, H., Brooks-Gordon, B., Campbell, R. Wild guesses and conflated meanings? Estimating the size of the sex worker population in Britain. *Critical Social Policy*, 29 (4), 1 November 2009: 703–19.

18. ACPO (Association of Chief Police Officers). Setting the Record: The Trafficking of Migrant Women in the England and Wales Off-Street Prostitution Sector. 2010. Available from: http://www.acpo.police. uk/documents/crime/2010/201008CRITMW01.pdf [accessed 5 November 2014].

19. PunterNet UK. Escort Directory, Reviews, and Forum. 2014. Available from: http://www.punternet.com/index.php [accessed 7 September 2014].

20. Millward, J. Dirty Words: A Probing Analysis of 5000 Call Girl Reviews. Available from: http://jonmillward.com/blog/attraction-dating/dirty-words-analysis-of-call-girl-reviews/ [accessed 7 September 2014].

21. Taylor, J. Punter Net prostitutes thank Harriet Harman for publicity boost. *The Independent*. Available from: http://www.independent.co.uk/news/uk/home-news/punter-net-prostitutes-thank-harriet-harman-for-publicity-boost-1796759.html [accessed 23 October 2014].

22. Groom, T. M., Nandwani, R. Characteristics of men who pay for sex: A UK sexual health clinic survey. *Sexually Transmitted Infections*, 82 (5), October 2006: 364–7.

23. MacShane, D. Tackling the trafficking myths. *The Guardian*, 16 November 2009. Available from: http://www.theguardian.com/commentisfree/2009/nov/16/trafficking-myths-sex-slavery [accessed 5 November 2014].

24. Gupta, R. The truth of trafficking. *The Guardian*, 2 April 2009. Available from: http://www.theguardian.com/commentisfree/2009/apr/02/women-sex-industry-trafficking-prostitution [accessed 23 October 2014].

25. Brooks-Gordon, B. Red mist obscures red light statistics. *The Guardian*, 3 April 2009. Available from: http://www.theguardian.com/commentisfree/2009/apr/03/prostitution-humantrafficking [accessed 23 October 2014].

26. Davies, N. Prostitution and trafficking: The anatomy of a moral panic. *The Guardian*, 20 October 2009. Available from: http://www.theguardian.com/uk/2009/oct/20/trafficking-numbers-women-exaggerated [accessed 5 November 2014].

27. National Crime Agency. United Kingdom Human Trafficking Centre National Referral Mechanism Statistics 2013. Available from: http://www.nationalcrimeagency.gov.uk/publications/139-national-referral-mechanism-statistics-2013/file [accessed 23 October 2014].

28. Lawson, R. The operation of the Contagious Diseases Acts among the troops in the United Kingdom, and men of the Royal Navy on the home station, from their introduction in 1864 to their ultimate repeal in 1886. *Journal of the Royal Statistical Society*, 54 (1), 1891: 31.

29. Howell, P. A private Contagious Diseases Act: Prostitution and public space in Victorian Cambridge. *Journal of Historical Geography*, 26 (3), July 2000: 376–402.

30. Lawson, The operation of the Contagious Diseases Acts.

31. Fisher, T. Josephine Butler: Feminism's neglected pioneer. *History Today*, 46, 1996: 32–8.

32. Public Health England. Sexually Transmitted Infections (STIs): Annual Data Tables. 2014. Available from: https://www.gov.uk/

government/statistics/sexually-transmitted-infections-stis-annual-data-tables [accessed 7 September 2014]. Public Health England. Sexually transmitted infections and chlamydia screening in England, 2013. *Health Protection Report,* 8 (24). Available from: https://www. gov.uk/government/uploads/system/uploads/attachment_data/ file/345181/Volume_8_number_24_hpr2414_AA_stis.pdf [accessed 7 September 2014].

33. Cox, D., Anderson, R., Johnson, A., Healy, M., Isham, V., Wilkie, A., et al. *Short-Term Prediction of HIV Infection and AIDS in England and Wales.* Her Majesty's Stationery Office, 1988.

34. Poor data hamper prediction on AIDS. *New Scientist.* 1988. Available from: http://books.google.co.uk/books?id=1M3e82yGmZMC&pg=P A14&lpg=PA14&dq=AIDS+projections+david+cox&source=bl&ots= FQIHuJ4_5N&sig=UOmMlx53kyZXbm3W6TYfZhwyFdQ&hl=en&s a=X&ei=VenQU8noIbPA7Aa264CABw&ved=0CDYQ6AEwAg#v=on epage&q=AIDS%20projections%20david%20cox&f=false [accessed 7 September 2014].

35. Public Health England. HIV in the United Kingdom: 2013 Report. Available from: https://www.gov.uk/government/publications/ hiv-in-the-united-kingdom [accessed 7 September 2014].

36. Boily, M.-C., Baggaley, R. F., Wang, L., Masse, B., White, R. G., Hayes, R. J., et al. Heterosexual risk of HIV-1 infection per sexual act: Systematic review and meta-analysis of observational studies. *Lancet Infectious Diseases,* 9 (2), February 2009: 118–29.

37. Downs, A. M., De Vincenzi, I. Probability of heterosexual transmission of HIV: Relationship to the number of unprotected sexual contacts. European Study Group in Heterosexual Transmission of HIV. *Journal of Acquired Immune Deficiciency Syndromes and Human Retrovirology,* 11 (4), 1 April 1996: 388–95.

38. Hooper, R. R., Reynolds, G. H., Jones, O. G., Zaidi, A., Wiesner, P. J., Latimer, K. P., et al. Cohort study of venereal disease, 1: The risk of gonorrhea transmission from infected women to men. *American Journal of Epidemiology,* 108 (2), 1 August 1978: 136–44.

39. Office for National Statistics. Crime in England and Wales, Year Ending March 2014. Available from: http://www.ons.gov.uk/ons/ rel/crime-stats/crime-statistics/period-ending-march-2014/stb-crime-stats.html [accessed 15 October 2014].

40. Office for National Statistics. UK Statistics Authority Assessment of Crime Statistics. 2014. Available from: http://www.ons.gov.uk/ons/ rel/crime-stats/crime-statistics/period-ending-september-2013/sty-uksa-assessment.html [5 November 2014].

41. Office for National Statistics. Crime Statistics, Focus on Violent Crime and Sexual Offences, 2012/13. Chapter 4, Intimate Personal

Violence and Partner Abuse. 2014. Available from: http://www.ons. gov.uk/ons/rel/crime-stats/crime-statistics/focus-on-violent-crime-and-sexual-offences--2012-13/rpt---chapter-4---intimate-personal-violence-and-partner-abuse.html [accessed 15 October 2014].

42. Home Office. An Overview of Sexual Offending in England and Wales. Available from: https://www.gov.uk/government/statistics/an-overview-of-sexual-offending-in-england-and-wales [accessed 15 October 2014].

43. Macdowall, W., Gibson, L. J., Tanton, C., Mercer, C. H., Lewis, R., Clifton, S., et al. Lifetime prevalence, associated factors, and circumstances of non-volitional sex in women and men in Britain: Findings from the third National Survey of Sexual Attitudes and Lifestyles (Natsal-3). *Lancet*, 382 (9907), 30 November 2013: 1845–55.

44. World Health Organization. Sexual Health. Available from: http://www.who.int/topics/sexual_health/en/ [accessed 5 November 2014].

45. Abrahams, N., Devries, K., Watts, C., Pallitto, C., Petzold, M., Shamu, S., et al. Worldwide prevalence of non-partner sexual violence: A systematic review. *Lancet*, 383 (9929), 16 May 2014: 1648–54.

46. IPPR. Young People, Sex and Relationships: The New Norms. 2014. Available from: http://www.ippr.org/publications/young-people-sex-and-relationships-the-new-norms [accessed 22 October 2014].

15: A Boy or a Girl

1. Office for National Statistics. Births in England and Wales, 2012. Office for National Statistics, 2013. Available from: http://www.ons.gov.uk/ons/rel/vsob1/birth-summary-tables--england-and-wales/2012/stb-births-in-england-and-wales-2012.html#tab-Live-births-within-marriage-civil-partnership [accessed 5 September 2014].

2. MacDorman, M., Kirmeyer, S., Wilson, E. Fetal and perinatal mortality, United States, 2006. *National Vital Statistics Reports*, 60 (8), August 2012: 1–22.

3. Graunt, J., Petty, S. W. *Collection of Yearly Bills of Mortality, from 1657 to 1758 Inclusive*. A. Miller, 1759.

4. Campbell, R. B. John Graunt, John Arbuthnot, and the human sex ratio. *Human Biology*, 73 (4), 2001: 605–10.

5. Arbuthnot J. An argument for Divine Providence, taken from the constant regularity observ'd in the births of both sexes. By Dr. John Arbuthnot, physitian in ordinary to Her Majesty, and Fellow of the College of Physitians and the Royal Society. *Philosophical Transactions*, 1 January 1710: 186–90.

6. Office for National Statistics. Births in England and Wales by Characteristics of Birth 2, 2012. Available from: http://www.ons.gov.uk/ons/rel/vsob1/characteristics-of-birth-2--england-and-wales/2012/sb-characteristics-of-birth-2.html [accessed 5 September 2014].
7. James, W. H. The variations of human sex ratio at birth during and after wars, and their potential explanations. *Journal of Theoretical Biology*, 257 (1), 7 March 2009: 116–23.
8. Mathews, T. J., Hamilton, B. E. Trend analysis of the sex ratio at birth in the United States. *National Vital Statistics Reports*, 53 (20), 14 June 2005: 1–17.
9. Hesketh, T., Xing, Z. W. Abnormal sex ratios in human populations: Causes and consequences. *Proceedings of the National Academy of Sciences*, 103 (36), 5 September 2006: 13271–5.
10. James, W. H. The human sex ratio. Part 1: A review of the literature. *Human Biology*, 1987: 721–52.
11. Catalano, R., Bruckner, T., Marks, A. R., Eskenazi, B. Exogenous shocks to the human sex ratio: The case of September 11, 2001 in New York City. *Human Reproduction*, 21 (12), December 2006: 3127–31.
12. Catalano, R., Bruckner, T., Anderson, E., Gould, J. B. Fetal death sex ratios: A test of the economic stress hypothesis. *International Journal of Epidemiology*, 34 (4), 1 August 2005: 944–8.
13. Kanazawa, S. Violent men have more sons: Further evidence for the generalized Trivers–Willard hypothesis (gTWH). *Journal of Theoretical Biology*, 239 (4), 21 April 2006: 450–9.
14. Kanazawa, S., Vandermassen, G. Engineers have more sons, nurses have more daughters: An evolutionary psychological extension of Baron-Cohen's extreme male brain theory of autism. *Journal of Theoretical Biology*, 233 (4), 21 April 2005: 589–99.
15. Kanazawa, S. Beautiful parents have more daughters: A further implication of the generalized Trivers–Willard hypothesis (gTWH). *Journal of Theoretical Biology*, 244 (1), 7 January 2007: 133–40.
16. Gelman, A. Letter to the editors regarding some papers of Dr. Satoshi Kanazawa. *Journal of Theoretical Biology*, 245 (3), 2007: 597–9. Gelman, A., Weakliem, D. Of beauty, sex and power. *American Scientist*, 97(4), 2009: 310–16.
17. James, W. H. The human sex ratio. Part 2: A hypothesis and a program of research. *Human Biology*, 1987: 873–900.
18. James, W. H., Valentine, J. A further note on the rises in sex ratio at birth during and just after the two world wars. *Journal of Theoretical Biology*, 363, 21 December 2014: 404–11.

19. Harlap, S. Gender of infants conceived on different days of the menstrual cycle. *New England Journal of Medicine*, 300 (26), 28 June 1979: 1445–8.

20. Hesketh and Xing, Abnormal sex ratios in human populations: Causes and consequences.

21. Department of Health. Birth Ratios in the United Kingdom. 2013. Available from: https://www.gov.uk/government/publications/gender-birth-ratios-in-the-uk [accessed 7 September 2014].

22. James, The human sex ratio. Part 1: A review of the literature.

Appendix: Natsal Methods

1. Erens, B., Phelps, A., Clifton, S., Mercer, C. H., Tanton, C., Hussey, D., et al. Methodology of the third British National Survey of Sexual Attitudes and Lifestyles (Natsal-3). *Sexually Transmitted Infections*, 90 (2), March 2014: 84–9.

2. Turner, H. A. Participation bias in AIDS-related telephone surveys: Results from the National AIDS Behavioral Survey (NABS) non-response study. *Journal of Sex Research*, 36 (1), 1 February 1999: 52–8.

3. Dunne, M. P., Martin, N. G., Bailey, J. M., Heath, A. C., Bucholz, K. K., Madden, P. A., et al. Participation bias in a sexuality survey: Psychological and behavioural characteristics of responders and non-responders. *International Journal of Epidemiology*, 26 (4), 1 August 1997: 844–54.

4. Copas, A. J., Johnson, A. M., Wadsworth, J. Assessing participation bias in a sexual behaviour survey: Implications for measuring HIV risk. *AIDS*, 11 (6), May 1997: 783–90.

5. Fenton, K. A., Johnson, A. M., McManus, S. , Erens, B. Measuring sexual behaviour: Methodological challenges in survey research. *Sexually Transmitted Infections*, 77 (2), 1 April 2001: 84–92.

6. Overy, C., Reynolds, L. A., Tansey, E. M. *History of the National Survey of Sexual Attitudes and Lifestyles: The Transcript of a Witness Seminar Held by the Wellcome Trust Centre for the History of Medicine at UCL, London, on 14 December 2009.* Queen Mary, University of London, 2011.

7. Mitchell, K., Wellings, K. *Talking about Sexual Health: Interviews with Young People and Health Professionals.* Health Education Authority, 1998.

8. Copas, A. J., Wellings, K., Erens, B., Mercer, C. H., McManus, S., Fenton, K. A., et al. The accuracy of reported sensitive sexual behaviour in Britain: Exploring the extent of change 1990–2000. *Sexually Transmitted Infections*, 78 (1), 1 February 2002: 26–30.

9. Turner, C. F., Ku, L., Rogers, S. M., Lindberg, L. D., Pleck, J. H., Sonenstein, F. L. Adolescent sexual behavior, drug use, and violence: Increased reporting with computer survey technology. *Science*, 280 (5365), 8 May 1998: 867–73.

10. Johnson, A. M., Copas, A. J., Erens, B., Mandalia, S., Fenton, K., Korovessis, C., et al. Effect of computer-assisted self-interviews on reporting of sexual HIV risk behaviours in a general population sample: A methodological experiment. *AIDS*, 15 (1), 5 January 2001: 111–15.

LIST OF
ILLUSTRATIONS

While every effort has been made to contact copyright-holders of illustrations, the author and publishers would be grateful for information about any illustrations where they have been unable to trace them, and would be glad to make amendments in further editions.

ACKNOWLEDGEMENTS

This book was conceived over a lunch with Andrew Franklin and Cecily Gayford from Profile Books and Kirty Topiwala from the Wellcome Collection. A book on statistics appeared a fine accompaniment to Wellcome's 'Institute of Sexology' exhibition, and the opportunity to write a book with such a good title was irresistible. But completing it was another matter, and I am extremely grateful for all the help and encouragement I got: in particular, Simon Szreter for introducing me to some fascinating history, Kaye Wellings and Anne Johnson from Natsal for talking to me at length, John Valentine and William James for helping me with sex ratios, Jon Millward for discussing his fascinating projects, Belinda Brooks-Gordon for insights into sex workers, Melissa Hines for the latest evidence on causes of sexual identity, Kim Horner for insights into what teenagers get up to in schools, and Philip Osment for reading and commenting on material.

When it came to data, Casey Copen from the National Survey of Family Growth provided recent reports, while Cath Mercer, Kyle Jones and Soazig Clifton from Natsal generously provided me with unpublished data, and also checked that what I said was not too misleading. I want to thank James Phillips for digging out wonderful historical nuggets from the British Library, and Cambridge University Library for being such a fine institution and providing me with some fairly 'top-shelf' books without a raised eyebrow, and also rapidly coming up with the most obscure official statistics. And I suppose I should thank my internet providers for not blocking me from searching the most inappropriate sites and material.

My agent Jonny Pegg looked after me as usual. Everyone at Well-come and Profile Books encouraged me when I flagged, but I am particularly indebted to Kirty Topiwala for her unending enthusiasm, and to Cecily Gayford for editing beyond the call of duty, and politely excising my coarse schoolboy remarks. Matthew Taylor sorted out my text with his usual meticulousness.

Matthew Taylor sorted out my text with his usual meticulousness, while Flora Willis, Penny Daniel and Drew Jerrison organised me in the kindest possible way.

The people at Nice and Serious have produced a fine app to go with the book. It was a pleasure working with them.

My colleagues in Cambridge were always supportive, and, of course, I am indebted to David Harding of Winton Capital Management for giving me the opportunity to indulge myself in this kind of project.

Finally I would like to thank my family, Kate, Kate and Rosie, for their continued encouragement, and for never once suggesting that perhaps this was not altogether a good idea.

Of course, all mistakes and misinterpretations remain my own.

INDEX